A Boolean Anthology

Selected writings of Mary Boole - on mathematical education

Compiled by D. G. Tahta and published by the Association of Teachers of Mathematics - 1972

Reprinted August 1980
Reprinted September 1987

CONTENTS

Mary Everest Boole (1832 - 1916)

George Boole (1815 - 1864)

Introduction

The collected works of Mary Boole amount to more than fifteen hundred pages. Among them is found a regular and passionate insistence on the real—but widely ignored—significance of her husband's work, and a vision of mathematical education that is powerfully relevant, but still unrealised, today.

This selection sets out to provide a brief introduction to her writings in order to stimulate a wider awareness of her thought. The numbers in brackets at the end of each extract are page number references to the collected works in the now out-of-print 1931 edition.

Mary Everest, later the wife of the mathematician George Boole, was born in 1832. Her father, a rector, was a friend of Babbage and Herschel who as students had initiated various reforms in mathematics at Cambridge. Her mother was the sister of a classics professor at a college in Cork, where Boole was to hold the chair of mathematics. When Mary was five her father became seriously ill and the family moved to France, so that he could be treated by Hahnemann, the founder of homeopathic medicine. Mary grew up bilingual. An early interest in mathematics was aroused by a teacher who gave her private lessons in arithmetic; many years later she wrote a charming account of this inspiration. Her father's health improved and the family returned to England when Mary was eleven.

Mary first met George Boole on a visit to her uncle in Cork. After the hard struggle of the early years, Boole was just beginning his great work. His masterpiece, *An investigation of the laws of thought*, was published in 1854 and was dedicated to Mary's uncle. Her father died the following year, and they married soon after. It was a successful marriage; though Boole was seventeen years her senior they were close companions, and Mary was able to share her husband's interests—she became a devoted disciple.

They had five daughters. The youngest was only six months old when Boole died of an attack of pneumonia in 1864. Mary was then 32 years old.

In the following year she took a post at Queen's College, London. Opened in 1847, this was the first college of higher education for women. Though unable to award degrees, it gave something more valuable, at least to some of the pupils; one of Mary's was to write later, "I thought we were being amused not taught. But after I left I found you had given us a power. We can think for ourselves, and find out what we want to know."

After a few years Mary left the college to become for a brief time the secretary of James Hinton, a writer on ethics and evolution as well as a skilful ear-surgeon, who had been a friend of her father. She was interested in the aspect of his work that most of his friends found unimportant, namely 'the art of thinking itself, which he, as well as George Boole, believed to be the true key to the physical and moral regeneration of mankind'.

In the twenty years after her husband's death Mary Boole supported and raised a family and read, corresponded and discussed over a wide range of themes and with a large circle of friends (some of whom are mentioned briefly in an appendix—see p. 75). Committed to spreading her husband's ideas about mathematics she also developed her own thoughts on education. She studied the works of Thomas Wedgwood, the son of the famous potter, and a friend of Darwin. She developed the ideas of the French logician and mystic, Gratry, and related them to those of her husband. At the same time she became more and more involved in spiritualism and theosophy with a curious amalgam of ideas from Hebrew ritual, Indian mysticism and Western science. At the age of 50 she embarked on a series of books and articles, publishing regularly up to the time of her death in 1916 at the age of 84.

Much of her writing has not been to twentieth century tastes though many contemporary young people have been rediscovering her themes. She was interested in the occult, homeopathy, vegetarianism, anti-vivisection—and love. But the practical common sense of her lesson notes, *Lectures on the logic of arithmetic* (1903, when she was 71), and the pioneering insights of *The preparation of the child for science* (1904) had their impact on progressive schools in England and the U.S.A. in the first decade of the 20th century. She invented curve-stitching and for many years cards marked for this purpose were known as Boole cards (see p. 35). Her first book, published in 1883 but written in the sixties, was a pioneer work on mental hygiene. In *the mathematical psychology of Gratry and Boole* (1897) and one of her last books, *The forging of passion into power* (1910), she showed an understanding of the unconscious far ahead of her time.

Most educationists today appear to have other things to write about than dowsing rods and so on. Certainly other things than love. Few achieve the clarity, the vision and the highly relevant insights of Mary Boole.

SELECTED WRITINGS ON LEARNING AND TEACHING

On influence

In all ordinary works on education, we may notice constant discussion about the particular mode in which influence should be brought to bear, whether by individual commands, unvarying rules, hopes of fears about consequences, appeals to affection, the contagion of example, or that more subtle form of influence called by the pious 'intercessory prayer,' and by modern Science 'suggestion,' or 'telepathy.' The whole discussion usually turns on the rival merits of the various modes of *bringing influence from without to bear on the pupil.* It seems assumed that it is always legitimate to exert influence. In mathematics, however, the main question kept in view is: When may the teacher exert influence?

For mathematical purposes, all influence from without, which induces the pupils to admit a principle as valid before his own unbiased reason recognises its truth, come under the same condemnation. (727)

A village schoolmaster

Monsieur Déplace was asked to come for two hours a day to teach us French and arithmetic. The only time he could spare was from six to eight in the morning, and those hours he spent with us, winter and summer, as long as we stayed at Poissy.

Monsieur Déplace is the hero of my idyll. I wish, though I know that the wish is vain, that I could convey any adequate impression of the way in which he enveloped my life with a protecting influence without the slightest interference with either my thoughts or my feelings. The influence was all the stronger because he showed no desire to gain influence; I was quite unconscious of it then and for many, many years afterwards. In those days the distinctions of rank were more sharply defined than they are now. The village schoolmaster was not supposed to take rank with gentlefolk, and though Monsieur Déplace himself was a gentleman in every sense of the word, he accepted the social position into which he had fallen (or risen?) in all seriousness.

I do not remember anything passing between us in the slightest degree resembling either a caress or an affectionate word. The relation between us was, I believe, entirely subconscious on my part, then, for a quarter of a centry later after I left him. I remember distinctly the first arithmetic lesson he gave me at home. Mother had been trying to teach me long division, but could not make me understand, chiefly, as I now know, because she herself did not understand. The difficulty was referred to Monsieur Déplace; then it disappeared as if by magic, and it was the last difficulty that I was ever in my life able to see in connection with arithmetic.

The next thing that I remember is a sum in compound proportion, or what used to be called 'double rule of three.' The master told me nothing, he asked me a succession of questions and made me write down each answer as I gave it, and then let me perceive that the answer to the complicated question on which we had started came out of its own accord in my own handwriting. (1514)

A university professor

She seemed so unlike the stuff of which mathematicians are made, that I one day said to her, "I cannot think mathematics is your true function."

"Oh, no!" she replied, "my *function* is to understand *whatever Professor Clifford wants to get understood.*" Of course every occultist will know what fatal magnetic interlocking must have been going on before such a thing could be said. To non-occultists I must explain that it was not at all the utterance of sentimental gush; the girl was, in the most straightforward manner, expressing her honest, though too limited, knowledge of the nature of her own mental activities. Professor Clifford, however, attributed the wonderful work done by his pupil to a personal vocation of her own for mathematics. The result was that her health broke down: then she took a fit of Evangelical piety—(i.e. found out too late that her function in life was to understand things which Professor Clifford did *not* particularly care to have understood), and died in the odour of sanctity—one of the countless victims of the cruel delusion of teachers who mistake the reflection of their own mental processes on the brain of an intuitional pupil for personal mental action in the pupil. In this case, as in many others, Nature seems to have avenged the wrong. From my own observation of Professor Clifford at classes of his which I attended, I venture to think that his premature decay, as well as much of his over-brilliant success, was due to a habit of recklessly over-magnetizing his pupils.

The Prophets of Palestine may not have known everything, but some of them did know that it is inadvisable to undermine one's own health and other people's for the sake of spreading one's special views and increasing one's influence, by random indulgence in the wonderful and awful delight of contact between intellect and intuition. (596)

Teacher lusts

The teacher (whether school-teacher, minister of religion, political leader, or head of a family) has a desire to make those under him conform themselves to his ideals. Nations could not be built up, nor children preserved from ruin, if some such desire did not exist and exert itself in some degree. But it has its gamut of lusts, very similar to those run down by the other faculties. First, the teacher wants to regulate the actions, conduct, and thoughts of other people in a way that does no obvious harm but is quite in excess both of normal rights and of practical necessity. Next, he wants to proselytise, convince, control, to arrest the spontaneous action of other minds, to an extent which ultimately defeats its own ends by making the pupils too feeble and automatic to carry on his teaching into the future with any vigour. Lastly, he acquires a sheer automatic lust for telling other people 'to don't,' for arresting spontaneous action in others in a way that destroys their power even to learn at the time what he is trying to teach them. What is wanted is that we should pull these three series tight so as to see their parallelism, and not go on fogging ourselves with any such foolish notion as that sex-passion is a lust of the flesh and teacher-lust a thing in itself pure and good, which may legitimately be indulged in to the uttermost.

Few teachers now are so conceited as not to know that they have a great deal to learn, and that their methods need revising and improving, but the majority are seeking for improved methods of doing more of what they are already doing a great deal too much of. The improvement which they most need is to be brought under conviction, to be made see their conduct, their aims, their whole attitude towards their pupils and their work, in the light reflected on them from those of the drunkard and the debauchee. (1412)

Qualities of a teacher

Constant recognition that man is as liable to error while his transcendental faculites are active as at any other time; and, moreover, that errors which otherwise would have been immediately detected and corrected, tend, if made at that time, to become fixed and to appear like solid truths, unless carefully checked by some well-selected mode of correction (e.g., the 'proving' of a sum).

Great reserve on the part of the teacher in even stating to pupils the special conclusions to which he has been led, lest he should arrest the normal exercise of their investigating faculties.

Specially stern self-control in the use of personal influence to convince or to persuade.

Modest reticence in the matter of courting applause by proclaiming any partial result before one has brought back into consciousness all those elements of a question which one had discarded from attention, in order to fix it for a time solely on some selected elements of the question.

Constantly accepting apparent disturbances of harmony and order as proofs that the true law of the case belongs to a higher order than that in which the investigator is at the time working. (637)

Authority

Three main symbols of authority have shared between them the attention of the world: the slave-driver's whip, the shepherd's crook, and the conductor's baton. A reasonable man should make up his mind which of the three he prefers: which he will submit to when it is his turn to submit and wield when the time comes for him to rule.

The slave-driver's whip has various modifications, conventionalised disguises: the sceptre, the mace, the truncheon, the cane. The appeal of them all alike is to immediate impressions on the senses. Their message is brutal but honest: "If you will obey my will, your sensations shall be more agreeable than they will be if you thwart my will."

The shepherd's crook is modified into a bishop's crozier. The functions of the two are similar: to keep the sheep from strenuous exercise in high altitudes, where their limbs grow fleet and their tissues tough; to keep off wolves who might dispute possession of any portion of the flock with the man who considers himself its rightful owner; to lead them into plentiful pasture, so as to make them fat and their flesh tender, and guide them cunningly at last into the yard of the slaughter-house. The whole system is simply one long deception—often of sentimental self-deception.

The conductor's baton exerts no control except during certain hours of practice and of performance. Once the appointed time has expired, every man is free to go where he likes and do as he chooses. He is freer (because more able) than he would have been without his occasional episodes of servitude, to play by himself whatever tune he chooses, or to enter into effective combinations with musicians not known in that conductor's orchestra.

Friends, under which symbol will you serve? And by which will you prefer to rule? (1413)

Haste is unchaste

Two methods of acquiring truth are possible to man; and these are, I think fitly typified (as the instinct of poets in all ages has led them to typify) by the two ways in which a man can expend his physical force. One sort of knowledge is technical, and can be acquired *a froid* from the outside, by deliberate study; the other is intuitional or actual, and should be gained only under certain conditions of pleasurable excitement. The former may safely be acquired at any convenient time, in any quantity, and in any accessible way, unless special circumstances seem to point to the contrary (short of such over-absorption as leaves insufficient time for rest and exercise). The latter kind of knowledge is, in itself and always, evil and impure, except at times when it is exceptionally right and sacred; such times being marked by the coincidence of three factors: maturity of the individual as to age; preparedness as to feeling; and suitability of circumstances.

It seems, indeed, that purity is in one sense the equivalent of patience; i.e. a willingness neither to precipitate in act, nor to suggest in words, nor to anticipate in thought, that for which the full time has not come. All haste is essentially unchaste. We need to make parents and teachers pure, before we can make children so. We must introduce into the popular mind the idea of chastity in education, in knowledge, in friendship, in hero-worship, in family ties, especially in religion. Many of the elements of our school and home life tend to destroy purity: the methods of teaching rendered necessary by the prevalent system of competition; the ambition of parents; 'odours, lights, wine, society, coffee, rivalry.' Many things tend to teach children that love and knowledge (I do not here mean the technical knowledge, but that which I have called the Actual or Intuitional) are things to be greedily sought rather than earned by patient waiting and long labour. The wife goes into hysterics if her husband is not in the humour to attend to her; the parents are aggrieved if the child is not 'fond' of them; each relative is offended if he or she is not a favourite. True knowledge is spoiled for the child by premature verbal statements of things which should dawn on his soul like a revelation; even the mighty mystery of contact with the Unseen is deflowered by what is called religious teaching. Surely all this tends to make the very idea of checking premature curiosity and greed about that which is to make children parents in their turn, an illogical one. (337)

Waiting

Oriental life differs from ours by being steeped in a curious patient *waitingness*, which tends towards fatalism and indolence. But when the Eastern patience is sharpened by contact with Western bustle, it translates itself into an exquisite chasteness, compared to which our methods seem coarse and, one might almost say, brutal. The Oriental teacher keeps his theories, his system, his classification, his knowledge, in the background, till the normal development of the pupil's mind requires that they should be brought into use. For instance, an English pupil learning a language from a moonshie in India is made to read an easy book; and to learn the declensions, conjugations, and irregularities only as fast as they occur in reading. A method, *apparently* the same as this, is adopted by many European teachers, and has been made the basis of certain systems, especially the so-called 'Mastery' system. But in these systems, the systematiser, or teacher, decided beforehand at what point the pupil shall be told of an inflection; whereas the very essence of the Oriental method is, that the teacher waits until the pupil's attention is attracted to the inflection by a difference in the look or sound of the word from its look or sound in a previous sentence. (360)

Withdrawal

The cultivation of the mathematical imagination depends chiefly on the child being put into the right attitude towards mathematical conceptions in his earliest years; and, after that, on the right use being made of certain nodes or critical points which occur here and there in each branch of mathematics, and which should be dealt with in a quite different manner from the rest of the course. These form the revelation crises of the pupil's mathematical history; when he draws near one of these *the human teacher should carefully withdraw his influence*, and simply watch to see that no seriously false impression is being formed. *His object should be to efface himself, his books, and his systems;* to draw aside a curtain from between the child and the process of discovery, and to leave the young soul alone with pure Truth. (919)

Training and teaching

The chief reason why courtesy, reverence, and a *certain kind* of docility are needful for those who would learn, is this: Truth is never received into the human mind without an admixture of conventions, of what may be called fictions. These fictions have to be introduced, used, and then withdrawn. It would be impossible to teach even so straightforward a subject as mathematics without the temporary use of statements which are not true to the nature of things. The history of a child who is learning mathematics, like that of human thought, is very much a record of alternate introduction of convenient fictions and subsequent analysis of their true nature. A class, like a public, tends at times to become groovy and mechanical; to mistake the accidental for the essential; to treat necessary aids to learning as if they were actual truths; to lose sight of the relative importance of various kinds of information, A class in Botany tends to forget that classification and terminology are not so much part of the life of plants as circulation and fertilization; a a class in Analytical Geometry forgets that the co-ordinates are no part of a curve. Just so, the reading public forgot, till Charles Darwin woke it up, that intermittence is no necessary part of Creative Action; although it is convenient for man, for purposes of classification, to *imagine* a series of intermittent acts. A student tends to such forgetfulness in proportion as he becomes mechanical in his work; the genius of a teacher is very much shown by the manner in which he contrives to arouse the interest and correct the errors of a class which is becoming too mechanical.

Theorists in education sometimes imagine that a good teacher should not allow the work of his class to become mechanical at all. A year or two of practical work in a school (especially with Examinations looming ahead) cures one of all such delusions. Education involves not only teaching, but also training. Training implies that work shall become mechanical; *teaching* involves preventing mechanicalness from reaching a degree fatal to progress. We must therefore allow much of the actual work to be done in a mechanical manner, without direct consciousness of its meaning; an intelligent teacher will occasionally rouse his pupils to full consciousness of what they are doing; and if he can do so without producing confusion, he may be complimented and his class congratulated. (468)

Education means the educing of faculty. Children need many things besides education; many things which can best be given—some of them can only be given—under a regime of orderly routine.

Among these good things are discipline and training. But it would be well to remember that, during the time that these other good things are going on, *education* itself is *not* going on. Education proper is given by rare and episodical occurrences, which give to those dormant faculties *which disciplinary routine is holding down and keeping quiet* opportunity and stimulus to start into active life. (1266)

Learning

In classical learning it is eminently desirable to secure that the right impression shall be made from the first; that wrong impression shall have as little time as possible to deepen themselves. We wish the child's eye and ear to become accustomed from the first to welcome the right and reject the wrong; we do not wish him to gain any habit of tolerating wrong impressions. If the child uses the nominative where he should use the accusative, and is not at once corrected, that is so much to the bad for his future progress; if he can be got not to remember a time when he used the word wrongly, that is much to the good. But in science there are, there can be, no absolutely right impressions; our minds are not big enough to grasp any natural as a whole; everything depends upon drawing right conclusions from combinations of impressions, each of which is in itself inadequate and partially misleading; and if the pupil is to be got into scientific methods, that is what he must be trained to do. And in order that he may learn to do it, it is sometimes necessary that each of a succession of 'wrong' impressions should have time to register itself on the brain and become part of its available stock. Such a statement may naturally convey to the scholastic mind trained in classical traditions an impression of disorderliness, but it does not imply disorder. Up-to-dateness is the cause of disorder; the haste, the greed to efface rapidly each partial impression, when we have nothing to substitute for it but some other impression equally partial, is not only unscientific but eminently disorderly. (882)

. . . . . The booliness of the method depends essentially on my not making any statement as to the nature of the connection between two groups of facts. The method is to set brains vibrating with the simultaneous consciousness of two groups of facts, *free from any hamper, from any opinions as to the nature of the connection between the two groups*, and start them investigating the nature of the connection, (1195)

The centre of gravity

For the first twelve years or so of a child's life it is impossible to guess what class of emotions or ideas will appeal most strongly to him later on. Therefore it is wiser to try to link the good habits which are being formed with any motive which seems to sway him specially; but rather to appeal to quite general ones: all the persons about him, especially those with whom he lives, wish him to be clean, punctual, polite, etc., and things will be made pleasant to himself it he does as they wish. This vague combination of altruism and egoism is a sufficient basis for the formation of character in childhood. During the period of adolescence the motives which will ultimately dominate begin to stir in the consciousness. Parents and teachers too often make the mistake of imagining that it comes within their function to determine by what motives a pupil shall be swayed. That question does not depend on any decision of theirs. As well might a hatching hen decide that the eggs on which she is sitting shall develop into partridges and not into ducks! What does depend, to some extent, on the hen's action, is, whether the development which is going on shall be full and harmonious or arrested and impotent; whether the ducks or partridges (as the case may be) shall have their limbs in good working order, or shall be lop-sided and helpless to carry out their own purposes. "If you train up a child in the way *he* should go," said an eminent psychologist of the last generation, "when he is old he will not depart from it. If he departs from the way in which you have tried to train him, it is because you have tried to train him in a way which *he* should not have gone, one in which Nature never intended him to go."

This is now perfectly acknowledged by all psychologists worth mentioning. They know quite well that the business of the teacher is to found, on a basis of motives which exist, habits which will be useful. The only correction of this formula which the present writer would venture to suggest is an addition. Found habits which will be useful on a basis of *such among existing motives as are likely to prove permanent;* or, in other words: Build up good habits on a basis within which falls the centre of gravity of the individual with whom you are dealing.

Now what do we mean by the centre of gravity of a character?

In any given individual, it will be found that some motives potent in phase A lose their hold in B; and some which are potent in B lose hold in A. Some desires and ambitions, which appeal strongly to him in A, seem to him in B unworthy and trivial; and some aspirations which stir profound emotion in B are judged by the waking discrimination of A to be too refined, too subtle, too altruistic, for the present stage of existence. But there is probably always a range of motives which appeal to the heart of the individual in both phases. Let us picture the whole range of motives which influence him in the A phase as represented by the left-hand circle in the diagram, and the whole range which influence him in the B phase as represented by the right-hand one; we shall think of motives operative only in A as Y-motives, those operative only in B as Z-motives, while X, the *quæsitum*, will stand for the (usually small) range of motives which stir his deeper emotions during his dream-moods, and yet are judged by his waking discrimination to be practical; and which are therefore able to stimulate him both to strenuous effort and to steady self-restraint. Within X lies the true centre of gravity of the individual's heart and conscience: "the soul that makes him one from first to last."

Habits formed while motives X are present to the mind are unlikely to collapse under any stress and strain of life, or even in the conditions respectively known as 'absence of mind' and as 'temporary insanity.'

It must be observed that the permanent motive is not always one predominantly present to consciousness. It is revealed rather by the quality of the individual's interest in other things than by his conscious interest in that thing. A girl may seem at one time devoted to music or some art, at another to philosophy or literature; but always under the dominant influence of some teacher, and stimulated by his approval. The desire for human approval is in this case likely to be a more permanent motive than either art or philosophy. Or a girl may be absorbed at one time in the study of music; at another time of history or literature; and what her soul is seeking through its various phases may be the Law of rhythmic beats of the Unseen. In such a case, the motive which may be relied on as a basis for ethical habit is the belief in the retaliations brought about by the recoil-power of destiny. A young person may seem at one time intensely pious, at another painfully worldly; the motive all through may be an artistic sentiment, which causes the imagination to be fascinated, at one time by the 'beauty of holiness,' at another by some artistic quality perceived in social life; the power to trust to, in such a case, is neither the influence of the Church nor that of the world, but the desire for the outward expression of harmonious Law.

It must be remembered that we are treating here of the choice, not of a professional career for the future man or woman, but of a basis for the formation of habits. The bread-winning profession of an individual should be of such a kind that he can be interested in it, but can also escape entirely from the thought of it when off-duty. The basis of ethical habit, on the contrary, should be some sentiment from which the individual never escapes; which is about his path and about his bed; something for the sake of which he is willing to work and the thought of which makes it easy to rest; which gives to life a meaning, and robs death of its sting.

In no cases, perhaps, can the subject of educational disappointment be studied more easily or with more profit than in those of little girls who show an early taste for what are called domestic pursuits: needlework, dusting, and the arrangement of the table of the room. Such precocious little housewives often become, in later life, the most hopelessly and helplessly undomesticated women. The reason of the failure would seem to be this: The mother assumes that her intelligent little helper has 'domesticated tastes'; whereas the child's orderly activity is probably due to its being, at a very early age, the only outlet for some nascent passion, either the love of approval, or the general instinct of kindness, or the desire to help whoever is greater and cleverer than herself. A little girl may be phenomenally clever at darning father's socks as long as father is the dominant influence of her life, and darning socks the only thing she can do for him. If she finds out too early that she can help him, or some indulgent uncle, by copying MSS., it is much to be feared that the needle will prove to be not her true vocation after all, unless special pains are taken to cultivate the taste for some years. As soon as the child goes to school, domestic work finds itself perhaps in violent conflict with the ruling passion; and the taste for it crumbles like a snapped 'Rupert's drop,' never to be restored. Domesticated tastes, in a woman, are the normal result of household work having been an outlet for the expression of X-motive between the ages of twelve and eighteen.

The question naturally occurs: how can parents and teachers find the centre of gravity of a young life, how discover the ultimately dominant motive? Very often they cannot do so: perhaps it is not best to probe in the matter too curiously. The important thing is that they should realise that there *is* a centre of gravity to each young life; and that it lies in the region where apparently conflicting passions mutually overlap. They will then try to link the habits most important to form, not with the passion which seems strongest at any given time, but with the greatest possible variety of motives, in the hope that the good habit will link itself with the quality of feeling, whatever that may be, which underlies all the various apparent motives.

In this matter, as in many others, the fact which it is most important for us to know is that of our own ignorance. We cannot know what is the centre of gravity of a young nature; let us then not act as if we knew. We know only that it lies at the meeting point of the character's extremes.

The power in which we must put our trust is not our own strength, but the Force which is given off where conflicting elements meet. Our Deliverer is not any eidolon which we have fashioned with the hands of our imagination, but the Unity who reveals Himself in the union of apparent opposites. (1390)

Hindrances to reform

Professor Perry has been raising a storm round the fact that students of engineering and electricity cannot use their mathematical knowledge to facilitate their study of real forces to anything like the extent which they would do if their knowledge of so much arithmetic or other mathematics as they have learned were real and vital. This has caused a revival of interest in a question which has been almost entirely ignored in England for a generation or two, though sixty years ago it received the earnest attention of some of our most learned scientific men; viz., *What are the conditions which favour a vital knowledge of mathematics?* It may surprise some readers to be told that those conditions are almost entirely moral and spiritual, rather than intellectual; but such has always been the verdict of the deepest students. Quite lately a teacher called on me to compare notes with me on the subject. I knew nothing of his views. . . . We found ourselves in entire agreement on three capital points:

1. The great hindrance to the vitalization of arithmetic teaching in past ages was the desire of the clergy to prevent any development of logical faculty which might lead the masses to doubt the evidence brought forward by them in favour of religious dogmas.
2. One great hindrance *now* is the desire of the classes in possession of the country's wealth to give themselves the emotional luxury of imagining themselves *Christians*, while retaining their hold on physical luxury in a world where the poor are starving.
3. Only *dead* mathematics can be taught where the attitude of *competition* prevails: *living* mathematics must always be a *communal possession.* (1009)

On competition

Children have a right also to a share in that still higher and purer delight, self-effacing communal Research. The opportunity for it should be provided in school. For if taken, under judicious supervision, by a group approximately equal in age and attainments, it is harmless and invigorating in itself, and the rhythmic beat between Altruism and Competition can be properly set up. Whereas, if the work of School and College is too prosaically and continuously competitive, the more promising pupils are often tempted to throw themselves eagerly into some group of adult thinkers, who overstrain and overstimulate in them the desire for communal self-effacement. Then comes the necessity for taking up the life of bread-winning; the desire for worldly-success reasserts itself. The consequences of this violent return to the worldly life, after too long a spell of Port-Royal-like Altruism, are not unfrequently disastrous to both moral stamina and intellectual power. It will be found a good rule to make all work which is properly *intellectual*, all which involves serious thought, either individual or communal; and to reverse competition for those portions of time which are devoted to the acquiring of skill, accuracy, and speed, by practice in what the pupils already understand perfectly. The stimulus of competition, when applied at an early age to real thought-processes, is injurious both to nerve-power and to scientific insight. (892)

Group work

The method is essentially communal not competitive. Hardly ever are there two children working on quite the same problem; therefore there can be no competition as to who answered best or quickest. At the end of the lesson, the whole class are in possession of some information which no member of it possessed before. It is not something told them by the teacher; it is often quite new to the teacher herself. The skill of the teacher is shown, not by the knowledge which she imparts, but by the manner in which she utilises the thinking power of the children for the purpose of finding out what she does not yet know. (1196)

The idea of testing woman's fitness for a true womanly life by competitive examination is not quite a modern one. King Lear tried it. (267)

Two elements in teaching

Suppose I am teaching, say, the process of multiplication. There are two things which the pupils can get out of my instruction: (A) skill in performing the operation of multiplication itself; and (B) a little of the power to find out for themselves how to do other arithmetical operations. Every process that I teach ought to be so taught as to add something to the pupil's chance of some day making out a rule for himself without the aid of a teacher.

If we add together all the A's of a child's arithmetical career, they constitute what I called, in a former article, the *body* of his arithmetical knowledge; if we sum up the B's, they constitute what is called its *life*. The sum of the combined A elements constitutes the ability to reckon the bulk or number of dead material and to keep accounts according to any system chosen by an employer. The sum of the B elements gives the extra power of bringing one's knowledge to bear in forming a sound judgment on problems connected with living forces—e.g., on the probable behaviour of a charge of electricity under certain conditions, or the probable honesty and stability of a certain commercial enterprise.

Now the A element in any mathematical lesson can be imparted while the class is alert and eager; the B element cannot be imparted except under the peculiar condition called by some mystic writers 'Silence in the soul' awaiting further Light.

The two states, the alert and the passive, alternate in any good educational *regime;* the alert phases being very much the longest, the passively recipient ones short but quite undisturbed.

But under stress of competition the passive mystic phases of study are being crowded out, The reason is that England is so saturated with the spirit of advertisement that, in any given committee, the majority are almost sure to be against the teaching of anything for which there is nothing to show at the *next forthcoming* examination. (1012)

On wholeness

But no intellectual culture is possible without some monistic action of the mind; teachers therefore interrupt A action every now and then to insist that pupils shall set up irregular short scraps of B action in the middle of A action. "Think, children, think, what does this remind you of?" "Don't you see that . . .?" They expect to receive immediate answers to questions sprung on the class in the middle of work: questions of a kind which should have been asked only at the end of the class, and left to be thought out at leisure and answered on some future day. This of course tends to weaken and disorganise the faculties of comparison and the grasp of general principles in the majority of pupils. But there are always a few 'naturally protected' monists in each generation whom teachers may to some extent harden and make inefficient, but cannot turn aside from their true function. Some of these become what are called philosophers: men who spend their lives in writing ponderous books to prove simple truths which ought to occur naturally to everybody. These men present to the outer consciousness of their readers and hearers ideas which should never come into consciousness except as the latter is informed by the unconscious or inner mind. The path of such a teacher is often marked by a broad track of nervous and moral wreckage: a phenomenon which seems to astonish many persons, but which appears to me as little to be wondered at as the digestion wreckage which would follow in the wake of a physician who should insist on feeding his patients with ready-made chyle. Few things are more important to mental health than that each individual should be able to generate within himself as much monistic philosophy as he is ready for, and should have the instinct to avoid absorbing any more than he is fit for. (1002)

The majority of teachers try, as I have said, to keep their pupils out of normal monistic phases, to an extent not sanctioned by any great psychologist. All sorts of excuses are made for this: the need to use up time in learning to know the outside world; the examinations looming ahead; pressure of competition, etc., etc. These excuses seem genuine, until one notices that the manner in which the practice of average educational practitioners differs from that of great educational authorities is reflected in other regions where no such motives can possibly have any sway. For instance, in a large lunatic asylum where I have stayed on visits, the doctors and nurses seemed to me as kind and conscientious a set of people as I ever saw; but, thinking of their conduct in the light of certain admitted principles of medical psychology, I often found myself wondering whether they or the patients were the more insane. (1001)

X and not-*X*

It seems to me that most discussions about Education are vitiated by one great flaw; Educationalists seldom recognise the proper inter-play of the two actions. They know that a child must be taught to see clearly the difference between some things; and they know that he must be taught to see the Unity between some other things; but they do not, in fact they dare not, realise that the process of sound thinking consists, emphatically, in seeing Unity where one has already perceived contrast. It is simply a farce—now that we know the law of man's reasoning organ—to dignify anything with the name of Education which does not provide constant opportunities for the human *reason* to say "*x* plus not-*x* form a Unity": and that about some *x* which the mere human perception has seen sharply contrasted with the not-*x*.

There is not the least practical difficulty in doing this. There is no reason why any intelligent man or woman should not learn, in a week, enough to make the test-equation of the thinking organ the norm of his or her business: as a teacher, or as a sick-nurse in cases where the nerves are disturbed; or for the judging of the precise amount and kind of delusion a patient may be suffering from. The only difficulty now in the way of its quite general use, consists in the fact that it leads to consequences startling to good people who claim that right to form judgments about things they haven't studied and don't understand. Many a teacher has begun to get a glimpse of what the Law really is; and shrunk back not daring to look any longer, knowing that, if he does, he must soon find himself doing things that will put him in conflict, not only with the Examiner, but, what is much worse, with that rampant Philistinism which goes about clutching to its breast a bundle of mutually contradictory opinions, which it persists in considering as sacred axioms, apparently under the impression that cut-and-dried falsehoods become a living moral sense, if one only hugs them close enough to one's heart. The Examiner is made the excuse for much dishonesty in teachers, for which he is in no way responsible. (568)

On circular 711

It may be right in regard to merely political matters to let the various parties do their worst, and quietly to mind one's own business; but it cannot be the duty of any mother, with a child at school, to be content with going through her domestic avocations, and to leave her child exposed, without special help, to a method of teaching which may, and in many cases will, leave the child capable of thinking any foolishness right, and with no power of correcting and controlling the natural flow of ideas which is the normal birthright of the young. After this specific warning, our readers who are fathers and mothers of children at school must blame only themselves if in ten years from now they find these same children growing into manhood and womanhood with their minds incapable of understanding the eternal verities. (1461)

SELECTED WRITINGS — SOME MATHEMATICS LESSONS

Pre-mathematics

Of arithmetic and algebra I shall say little here, as they are treated in another chapter. Only one point I will lay stress on. Many a life of intellectual muddle and intellectual dishonesty begins at the point where some teacher explains the rule for Greatest Common Measure to a child who has not had the proper basis of sub-conscious knowledge laid in actual experiences. Therefore, if you value your child's future clearness in science, trust no teacher to tell him anything about G.C.M. or L.C.M. till you have ascertained that he is able to find, easily and accurately, by means of compasses, the longest length that will repeat exactly into each of two unequal given lengths, and the shortest length into which each of two given unequal lengths will fit.

We now come to the subject of geometry, the condition of which affords, it seems to me, a standing warning against directing educational care too exclusively to the conscious mind, and neglecting to provide food for unconscious mental action.

There seems to be evidence that in ancient times all people in good society were expected to know simple truths about geometric forms in the same way as we all know simple facts in natural history. The elementary properties of the triangle, parallelogram, circle, ellipse, and spiral seem to have been familiar to ordinary people. They were not expected to know much about geometry, but they were expected to have and to use ordinary faculties of observation on facts within every one's reach. Euclid, was in his day, a sort of Darwin of geometry. He wrote not a geometry for beginners, but a book about the logical concatenation of geometric facts for men already geometers; just as our Darwin wrote a book about the concatenation of biologic forms for people already biologists, to the extent at least of knowing that horses prance and dogs bark and wag tails; that worms creep and birds fly; that some flowers have scent; that some fruits are sweet and others are sour.

Euclid's book was a type and model of all that a good book on logical concatenation should be. The use which was made of it till lately is the type and model of how such a book should not be used. Teachers assumed that the excellence of the book gave them the right to use it in defiance of all the laws of psychology. The result of such misuse is always the same: *loss of natural instinct*. Textbooks are written expressly on purpose to inform the *consciousness*. A good textbook should explain everything step by step, and should assume nothing which it does not actually state. Euclid does this in perfection. He wrote, as I have said, for men for whom the words triangle, circle, parallelogram were already charged with associations; and he gave definitions intended for the purpose, not of telling something fresh, but of clearing up and settling conceptions which were hazy from long familiarity. . . .

Now when it became customary to give to boys of ten or twelve what Euclid wrote for grown men, that was not far wrong; boys now can quite well assimilate what was grown-up food two thousand years ago. But if children of twelve are to learn what Euclid wrote for advanced men, children of three should be acquiring the subconscious physical experiences which lads in Greece picked up in the course of nature and by the accidental help of architecture and statuary. This precaution our grandfathers entirely omitted. The effect was somewhat similar to that which would be produced if it ever became the fashion to make children learn theoretic natural history from books illustrated by flat diagrams, before allowing them to see any real animal or plant. *Europe has lost geometric instinct and the habit of geometric observation.* All of us at this time are in a condition of artificially paralysed geometric faculty; and now the aim and study of all true mathematicians is to restore the vitality of geometric instinct. . . .

Lastly—and this is probably the most important preparation for future living comprehension of mathematical ideas—there is the cultivation of the geometric imagination. At the same age at which the child begins to realize that a tadpole grows into a frog, a boy into a man, a seedling into a flowering plant, let him have the opportunity of watching also how one geometrical type-form grows out of, or flows into, another. A common night-light placed in the bottom of a deep round jar in a dark room throws on a sheet of cardboard held over it patterns of conic-sections, which pass into each other as you change the position of the cardboard. Children very early learn to love watching figures thrown in light; and there is no age at which this amusement can hurt them, provided that the motion is slow, and that no one excites them by trying to explain things. A variety of other methods for training the geometric imagination at a later stage will be dealt with in a future chapter. (904)

Lesson on equivalence

12 buttons are a dozen. 12 pence are a shilling

Are those statements true?

Are they both true?

Are they always true?

Are they true in the same sense?

Let us see.

Suppose you go into a shop and say: "I want a dozen of those buttons, please," and a friend says: "And I will take twelve of the same buttons, please," do your two purchases look alike? Would the two lots weigh the same? Would the owner of one of the lots be any the worse or the better off if the parcels were changed by accident? If the twelve buttons were sewn to your coat, would any one know that they were not the dozen? If we wanted to play a game with twelve counters and had no proper counters, we might use the twelve buttons for counters; would the dozen do instead? Yes, just as well. The dozen is twelve, and twelve is a dozen; and for every purpose for which one could be used, the other would do just as well.

You say twelve pence are a shilling. Do they look like a shilling? Are they the same colour, size, weight? If I wanted things to use instead of counters, I might use twelve pennies: would a shilling do instead? No. Sometimes in cooking, if we have not small weights, we use a coin as a weight; we might be told in a cookery book to take the weight of a sixpence or shilling of carbonate of soda or of some spice. How would it be if we used twelve pence instead?

Twelve pence are not a shilling, not in any way like a shilling. Why do you say they are a shilling? They are of the same value. Value of what? Not for counters, nor for weighing things.

Twelve pence are of the same value as a shilling when we want to buy things. Yes, now we have it right; twelve pence are not a shilling, and cannot be used instead of a shilling for any real use of the things themselves. But the chief purpose of coins is to exchange for other things; and, *for that purpose* twelve pence are of the same value as a shilling . . . (827)

Lesson on numeration

What is this 1? And this, 2? And this, 3? (and so on), and this, 9?

Now I want to write ten: how shall I do it?

Put 1 and 0.

But what has ten done to be different from the rest?

Why should it have two signs instead of one like its neighbours? and why does it take signs belonging to its neighbours, instead of having one of its own?

Did ten ask to have two signs? Did it wish to have two? No; then why do we give it two?

I once asked a young friend of mine why he did something in his sum; and he answered: "My reason is that I was told to do it at school; but I know I ought to have another reason, and I know I haven't." I thought that was a sensible answer. It applied to most things in arithmetic . . . (818)

Lesson on zero

Shut eyes, etc. Make a mind-picture:- Me lifting the chalk to the black-board. I make one stroke and then put my hand down. I do this action three times; how many strokes will be on the board? If, instead of making one stroke on the board, I made two and put my hand down; how many strokes would be on the board when I had done the action once? Twice? Three times? Four times? Before I had done it at all?

Open eyes, sit up. Shut eyes, etc. Make a mind picture:- A clean black-board, me holding the chalk and then putting it down, without touching the board; what would be on the board? Nothing. Now make another picture:- Me making a stroke. Now I rub the stroke out. What is on the board that you now see in your mind? Nothing. So if I do nothing, or if I make a stroke and rub it out, the result is the same as far as the board is concerned. Were the two ways of getting it the same? No.

That was a mind-picture black-board. Now open eyes and look at the real one. Here is a quite clean board; I make a stroke; I rub it out. Is it really quite as clean as when it has been cleaned on purpose for class? No. What do you see on it? A smudge.

Suppose I lifted the chalk to the board, and made a dot and no stroke at all; how many strokes would be on the board when I had done the action three times? Four times? Six times? Nine times? Once? Before it had done it? . . . (849)

Lesson on triangles

At Prop. XLVII (Pythagoras' theorem) it becomes even more important that the lesson should be as little as possible connected with instruction from a human teacher. It is, or should be, one of the most momentous crises of the mental life.

The question is posed thus: "You know that any two sides of a triangle are greater than the third side. How much greater depends, you know, on the angle between the two. Now suppose we leave out of reckoning all triangles except those which have a right angle. Is there any way in which we could know the length of the hypotenuse by knowing the lengths of the other sides?"

The child is left to grope and fumble over that question for a short time every day—for perhaps a week; so as to get the sense of impossibility, of helplessness, well worked into his consciousness; because it is important that he should get an idea of the true relation of mathematical method to the sense of *helplessness* generated by the limitation of direct human faculty. You then tell him to make a right angle by placing three of his square counters or blocks that go to make that. He will find that he needs five. He will then perhaps suspect that the law of the relation is contained in some idea of sequence of numbers. Let him then try four, five and six; he will find that six does not make the hypotenuse long enough. After trying various measurements for some days, you again tell him to make the triangle whose sides are three and four and the hypotenuse five. Tell him then to build up the complete square on each side. Then tell him to count the blocks in each of his squares. Then tell him to add together the numbers of blocks in the two small squares. And then, if you can, have the tact and wisdom to be silent and let him think. Say no more of geometry that day. Next day let him build the hollow triangle and solid squares of 3, 4, 5 again. Then other right-angled triangles, such as those with sides 5, 12, 13; 8, 15, 17; etc. Use no unnecessary words. (928)

Lesson on toads

A teacher took a class of little children into the wood, caught a toad and put him into the middle of the group, saying: "Our lesson today is to be about toads. But I know nothing about toads; someone else will give you the lesson."

One child said: "I suppose God will give us the lesson." Another interposed: "*I* think the toad will." (1307)

Lesson on infinity

Let us go over the ground again. Suppose there is a cake on the table. How many children can go through the room without the cake being all eaten up?

Well, that depends on two things: the size of the cake, and the share which each child eats. If the cake weighs two pounds, and each child eats two ounces, it will be all eaten up when sixteen children have gone through the room. If the cake weighs only one pound, it will be eaten up when eight children have gone through the room. But if each child eats only one ounce, then again sixteen children will have to go through the room before the cake is eaten up, and so on. Many questions could be asked, all depending on the size of the cake and the size of each child's share.

All this time you are tied to an hypothesis that the children eat cake (more or less).

But now suppose we are freed from that hypothesis. Suppose no cake is given to the children. How many can pass through the room before it is all eaten up?

The answer to that is: "An infinite number." Infinity does not mean any particular number, or a very large number. It means a loosened chain, a discarded hypothesis, escape from the rule we were working under. Something else, not the size of the cake determines the number of children. Infinity does not mean that there are enough children in the world now to go on passing through the room *for ever*, but that the number of children who pass through the room, now that the share of each child is 0 (zero), will have to be determined by the number of children that there are in the school, or the parish, or whatever it is that the children are supposed to come from; *and not by the size of the cake.* The size of the cake no longer has anything to do with answering the question: "How many children can pass through the room before the cake is all eaten?" (1258)

Lesson on curves

Try to fancy that this black-board is a field. It has a brick wall all round it, with no opening except at A, where there is a gate.

At Z there is a rabbit hole. A rabbit came out into the field at Z and wandered about till he came to here (*write the figure* 1), where he found something he liked to eat. After a little while, a dog came in at the gate A; the rabbit caught sight of him, and directly afterwards the dog caught sight of the rabbit. Now let us try if we can make out what happened. First we must try to think what the animals would each like to have happen; the rabbit saw the dog first; what do you think he wished? To get away. Perhaps his first idea is to run *straight away* from the dog. But he can't; the wall prevents him. What will he do next?

Now tell me, what do all these lines represent? The line 1 to Z represents the path which the rabbit would like to go along in one jump, and does take in eighteen jumps. The lines A 1, B 2, C 3, and all the other straight lines, each represent a line that the dog at some moment wished to jump along; he jumped along a bit of one, and then changed his mind and jumped a bit of the next, and so on. We drew all those straight lines; you saw me draw them by the ruler, did you not?

But here is a curved line A to Z. Who drew that? I drew no line except straight ones by the ruler. Look at it well. Make sure that you see it and all the lines on the black-board.

Now sit slack, shut eyes, and think what the curved line is and how it came. Open eyes and sit up. What is the curved line? The path which the dog really ran, when at each step he meant only to go down some straight line. (868)

You are going to begin learning about curves. It will help you to keep out of many muddles, if you will try to remember that, when you see a curve in a book or on paper, it represents some real form or movement; or something more or less like a real form or movement. It may be simpler than the real path or shape; but it is *only* simpler; it is really more or less like something meant to be real.

But when you see straight lines, they are seldom meant for anything real. A straight line represents either a path that one force alone *would* have taken something along if no other force had interfered; or else it is just put in for convenience, to measure by. You have perhaps seen tailors' fashion books, with directions for taking measures. You see a picture of a man with a coat on, and straight lines drawn across the shoulders or bust. You would be dreadfully puzzled if you thought of those lines as parts of the picture; because no coat has lines across the shoulders or bust; but you know they are meant, not for seams in the coat, but for a measuring-tape supposed to be stretched across the man in order to measure the width of his coat. I have known children puzzled out of their wits, and never able to understand their Geometry for years of their schooltime, because they mistook straight lines for real parts of some curved thing. (870)

Curve stitching

The use of the single sewing cards is to provide children in the kindergarten with the means of finding out the exact nature of the relation between one dimension and two.

There is another set of sewing cards which is made by laying two cards side by side on the table and pasting a tape over the crack between them. This tape forms a hinge. You can lay one card flat and stand the other edgeways upright, and lace patterns between them from one to the others.

The use of this part of the method is to provide girls in the higher forms with a means of learning the relation between two dimensions and three.

There is another set of models, the use of which is to provide people who have left school with a means of learning the relation between three dimensions and four.

The use of the books which are signed George Boole or Mary Everest Boole is to provide reasonable people, who have learned the logic of algebra conscientiously, with a means of teaching themselves the relations between n dimensions and $n + 1$ dimensions, whatever number n may be. (1247)

The beauty of some of the designs is unquestionable; and there can be no second opinion about the value of the method, as training, from the point of view of geometry as well as from that of art. What is not quite so obvious at first sight is its bearing on the training of the unconscious mind for science. Without the slightest intellectual strain it puts the children through that normal sequence of orderly attention to classification and detail, interspersed with nodal points of synthesis, which may be called the very breathing-rhythm of the scientific discoverer.

But to make this exercise of any use there must be no copying from diagrams; the value of it depends on the child evoking a curve, watching it growing, under his fingers, from mere obedience to a law. (906)

I have seen in print disparaging comments on play-methods of introducing children to arithmetical ideas. Such criticisms seem to be founded on an erroneous assumption that the play-methods are intended to teach something about arithmetical procedure, for which purpose they are, as the critics perceive, ludicrously inadequate. But these methods have no relation to the teaching of anything: their use is to give to the infant brain a start on a line of

SERIES II. **BOOLE CURVE-SEWING CARDS.** No. 2.

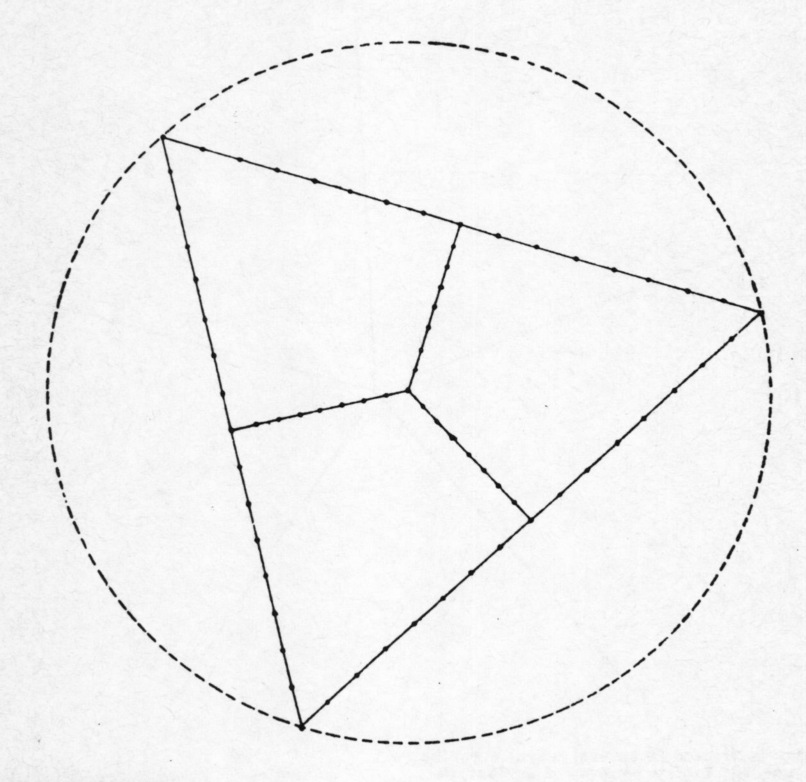

To be treated according to fancy.
Circle may be used if wished.

Knots and finishings on this side.

G. P. & S., LTD. LONDON.

Series II. **BOOLE CURVE-SEWING CARDS.** No. 4.

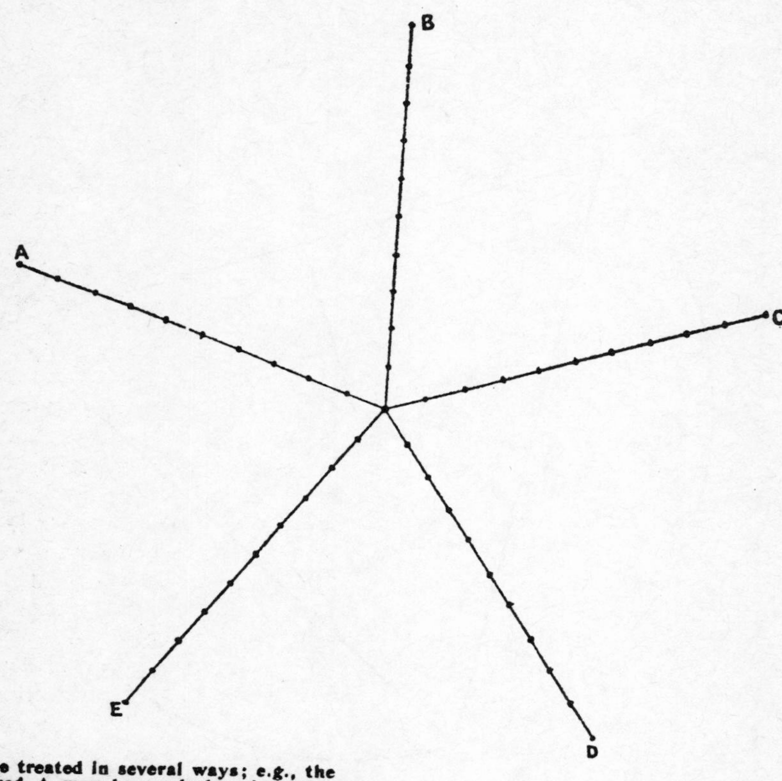

This may be treated in several ways; e.g., the line marked A may be worked against that marked C, C against E, E against B, B against D, and D against A. Or E against D, D against C, etc. Or A against B, B against D, D against A, and then C against E, etc.

Knots and finishings on this side.

G. P. & S., Ltd. LONDON.

development such that, in after years, when the intellectual teaching begins, it will be capable of receiving that teaching. I should be reluctant to insist on any opinion contrary to that of more experienced teachers than myself, *in any matter about which they have the experience*. But the majority of professional teachers, at present, on their own confession, have no idea on what depends the receptivity of the child when the teaching age for the subject has begun. Parents also are still so unawake to this that they confuse preparation for a subject with premature teaching of the subject; and do great harm by such premature teaching. Many teachers strongly deprecate amateurish attempts to instruct children before the proper time; and in this I entirely agree with them. A concrete instance may help to make clear in what preparation for a subject consists. In my young days cards of different shapes were sold in pairs, in fancy shops, for making needle-books and pin-cushions. The cards were intended to be painted on; and there was a row of holes round the edge by which twin cards were to be sewn together. As I could not paint, it got itself somehow suggested to me that I might decorate the cards by lacing silk threads across the blank spaces by means of the holes. When I was tired of so lacing that the threads crossed in the centre and covered the whole card, it occurred to me to vary the amusement by passing the thread from each hole to one not exactly opposite to it, thus leaving a space in the middle. I can feel now the delight with which I discovered that the little blank space so left in the middle of the card was bounded by a symmetrical curve made up of a tiny bit of each of my straight silk lines; that its shape depended upon, without being the same as, the outline of the card, and that I could modify it by altering the distance of the down-stitch from the up-stitch immediately preceding. As the practical art of sewing perforated card was already quite familiar to me, my brain was free to receive as a seed the discovery I had made, and to let it grow naturally; all the more because no one spoke to me then of tangents, or tried to teach me any algebraic geometry, till some years had elapsed. Therefore, when I did begin to learn artificially about tangents, the teacher was not obliged to put cuttings into raw soil; he found ready a good strong wild stock of living interest in the relation between a curve and the straight lines which generate it, on to which he was able to graft the new knowledge. The teacher came, not as an outsider thrusting on me the knowledge of something unfamiliar and strange; but as a brother-seer more advanced than myself, who could show me how to make further progress on a path which I had already entered with delight. On such accidents as this of my card-sewing depends, I think, much of those special receptivities for certain subjects, quite distinct from great power, which puzzle psychologists. When we understand better how they originate, they will no longer depend on accident, and we shall more often be able to produce them at will. But let me repeat once more: the receptivity cannot be generated by early teaching of a subject mixed up with the use of its appropriate technical machinery; but only by suggesting the new ideas by means of objects already familiar to the child's eye and touch, or of processes become habitual and automatic by use in connexion with a familiar idea. Unless this and a few other principles of normal sequence, admitted by all medical psychologists, are incorporated into the basis of our educational system, it is to be feared that more improved "methods" will increase the danger of abnormal sequences, and will turn out to be improved methods of injuring nerve-stamina, intellectual grasp, and moral stability. (914)

Mind pictures

The pauses of sitting with slack muscles and taking slow quiet breaths, to make mind-pictures, is not intended to supersede the usual permitted "intervals" between lessons; they are an integral part of the Logic lesson itself. They, as well as the rest of the scheme here suggested, form, it is believed, the first attempt to adapt to elementary education here the magnificent method of study described in Gratry's *Logique.* A certain element of normal mental life, essential to sound intellectual development, and which can only be received during suspension of the active perceptions of intellect, is being driven out by modern educational grind. This tends to produce a sort of mental and moral rickets, analogous to the physical disease induced by a diet deficient in the bone-forming element. Teachers who have studied any sort of medical psychology bitterly lament that the requirements of existing systems force them to inflict this grievous wrong upon the rising generation. In the present volume, an attempt is made to re-introduce at least a small share of the skeleton-forming elements of mental life, in a manner which does not necessitate disturbing existing arrangements. The author has been kindly assisted in this matter by Mrs. Archer.

English children have now to be taught to relax, to breathe, to see. It is probable that the cultivated art of relaxation will ultimately prove more satisfactory than the mere following of natural impulse to relax. (813)

The cultivation of the mathematical imagination should include not only its development but its orderly and systematic exercise. A child is too often made to pass from a particular case which suggests a law to other particular cases which require an application of it, without a sufficient amount of intermediate drill in analysis of the law and in tracing out its results exhaustively. Between the time when a child handles an actual cube, cuts sections, etc., and the time when he comes, among his ordinary geometrical exercises, to problems requiring him to draw the elevation of a cube cut in some particular way, there is a period when he finds it useful, and very delightful, to go through a set of processes in imagination and to express them in his own words. "I think of a cube. I think of slicing it, beginning at one angle. I get a point, then a little triangle, then bigger triangles. The corners are cut off, making a hexagon with unequal sides. The odd sides increase; the hexagon now has all its sides equal. The sides which were longer before are now shorter than the others. They disappear and the hexagon has become a triangle again, but with its angles in the other directions. The triangle becomes smaller and smaller. It becomes a point. It disappears." (929)

That is enough for today. Sit slack, shut your eyes, and rest before you go to the next class. I am going to give you two mind-pictures:

A little boy was so excited about going to a treat that he would keep talking about it, and would not tie his shoe-laces properly. On the way downstairs his lace came untied; he stepped on it and fell and sprained his ankle, so he could not go to the treat after all.

A servant was asked to get some children dressed to go out. She got them tidy, and their boots nicely cleaned and tied on; the children felt impatient because they were in a hurry to go out; but they knew they could not go till they were dressed, so they were good and quiet. When they were ready, they thought they were to go. But the nurse grumbled and said: "Now you are dressed and all tidy and clean, I don't want you to go out, for fear you should tumble your hair or get your boots dirty. I was told to dress you, and I have dressed you: that was the important thing; that was my real duty. Going out is all nonsense. Sit still all the afternoon, and look at your nice clean boots."

Try to fancy how these children would feel, and what they would think. Try for a minute or two to fancy yourselves in the place of those children, and to think what you would feel like (932).

The spiral

The Spiral has many forms. Perhaps the subtlest and most profoundly instructive is that traced by whirling wind in the dust of the road. Another is the path of a planet in space. One of the most important is the serpent; the old symbol of wickedness, and of wisdom, of disease and of its cure.

The modern student will better understand what the serpent meant to the ancients if he will first familiarise himself with other forms of spiral more accessible in civilised life. The simplest, the best to begin studying with, is the ordinary spiral wire. Let the student when he has half an hour to spare fetch the corkscrew.

If, having read so far, he smiles superior, imagining that no spiritual instruction can possibly be got out of so humble a domestic implement, he has still a good deal to learn before he can know the elements of the science of notation. The true student will think it well worth while to spend a dreamy half hour in finding out whether there is or is not anything to be learned from a corkscrew.

Stand the corkscrew up on end on the table; settle into a comfortable position, take a few long, easy breaths, and look at the screw, with the bodily eyes half-closed, and those of the imagination wide-open.

Imagine the screw prolonged to reach the ceiling. Imagine a crowd of microscopic creatures creeping up the screw. Their destiny is to rise from the table towards the ceiling. They have no road along which to ascend, except by the screw wire; and no consciousness of motion except in horizontal direction; e.g. they can recognise north, south, east, west, north-east, etc., but are not conscious of 'up' or 'down'. The inspired among them know however, that along the wire is their true destiny; the ceiling exerts a magnetism on them all, which the inspired among them feel, and which they know to be prophetic.

What next? Nothing much. Imagine the discussions that would take place: Is the true direction of progress north, south, east or west? Is there no such thing as right or wrong? Surely we must draw the line somewhere? If going north was right yesterday, going south cannot be right today? Why can we not be consistent. Why not decide, once for all, in what direction we mean to go? If our fathers found out twenty years ago, that going east was wrong, surely it cannot be right for us to go east now? And so on and so on. Only those can truly interpret the Present who understand the doubts and difficulties of the Past, because they have consciously felt the magnetism of the Future. (1310)

SELECTED WRITINGS ON THE NATURE OF MATHEMATICS

On mathematical certainty

One singular characteristic of mathematics is the automatic power of self-correction which the mind possesses in relation to it. No such thing is possible as the existence of either a persistent or a widely spread error, or a serious difference of opinion, as to the result of a calculation. And this is not due to any special immunity from error which we have in connection with the subjects called mathematical. Every schoolboy knows that it is as possible to make a mistake in a sum as in any other exercise; and the greatest mathematicians occasionally make mistakes in calculations. But there seems to be some mysterious court of arbitration within man which detects mathematical error. Any two persons who have come to different conclusions as to the result of a calculation can find out which of them was mistaken, without appealing to any external authority. Nor need they examine any outer facts; they need look at nothing outside of themselves except the slates or papers *which contain the register of their own successive mental acts*. They judge between themselves, and judge infallibly. When one child only has done a sum, he may need to look at the "answer in the book"; if his result does not correspond with that answer, he thinks that his processes must have been wrong somewhere. But when two persons have made the same calculation with differing results, there is no need to compare the results with the answer in a book; they revise and compare their own mental processes as registered by themselves. (703)

The sentimental people who assert that everything in Arithmetic can be "proved" to children have, usually, no idea of what rigid proof means; it is not necessary that the child should see the evidence for every hypothesis on which he works; what is necessary to mental health is a clear understanding of what constitutes evidence, and the power to distinguish between what is, and what is not, proved. (809)

The tendency in science classes to an over-cultivation of mere monkey-destructiveness has been largely fostered by the claim of certain faddists that children shall be told nothing, but shall "find out everything for themselves." What science really does claim in this matter is that a clear line of demarcation shall be preserved between *what the individual has observed* and *what he has learned at second hand*. The claim that children shall find out everything and be told nothing is palpably absurd. (889)

When we desire to know how a particular flower is fertilised, how a certain bird builds its nest, how an operation is performed by those skilful in it, that knowledge must come to us from the outside: we must see the facts, or hear lectures, or read books. Or, if we wish to know what the learned suppose to be the interpretation of certain phenomena, we must get the learned to tell us, either by speech or by writing. But if we wish to "think out for ourselves" the meaning of phenomena—i.e., to receive, without human instruction, new light about facts we already know—then we must keep mathematical order in our sequence of mental operations, or a delusion may come upon us, and we shall be likely to believe a lie, and perhaps to fix it on the texture of our thinking-machinery. (884)

Mathematical certainty depends not on the subject matter of our investigation but upon three conditions. The first is a constant recognition of the limits of our own knowledge and fact of our own ignorance. The second is reverence for the As-Yet-Unknown. The third is absolute fearlessness in meeting the *reductio ad absurdum*. In mathematics we are always delighted when we come to any such conclusions as $2 + 3 = 7$. We feel that we have absolutely cleared out of the way one among the several possible hypotheses, and are ready to try another. (1237)

On mathematical methods

But the discipline in mathematics, however strict, is entirely positive. There is in mathematics no "Thou shalt not." Provided that the necessary amount of practice in right methods be secured, during school hours, no one watches to see that wrong ones are avoided at other times. No mathematician worth mentioning has any objection to his pupils trying experiments in bad methods to any extent they please in their spare time. It never occurs to any one that boys are likely to form bad habits, or be the less teachable in class, because of any vagaries they may practise out of it. Mathematical method is a thing about which there exists no *esprit de corps*, no public opinion, nothing resembling either a canon of taste or an ethical code; and when conventions are imposed on the young by authority, it is with the distinct understanding, from the first, that any one who likes will be at liberty to alter the method in his spare hours; that he will be even applauded for doing so if he is fortunate enough to hit upon a more convenient one. In no department of study is reverence for precursors more cultivated than in mathematics; but that reverence is entirely dissevered from the notion that ancient methods should be adhered to; it is connected rather with a feeling of gratitude to those who, by inventing clumsy and imperfect processes, have helped successors to better ones. (729)

But when a principle has been admitted as valid, there is usually more than one possible way of carrying it out in any given case. The pupil ought to be able to see that the methods are essentially equivalent; but he is not at first capable of judging which of the several possible alternative methods is the most practically convenient. He will not be able to do so for many years to come; and meantime he ought to be forming the habit of some convenient method. It is here, therefore, that the principle of authority comes in (using the word authority in the sense indicated above, of any force or motive brought to bear on the pupil from the outside to induce him to do this rather than that).

It should be understood from the first that no such thing exists as *a* right method of performing any operation in elementary mathematics; because all rightness, and I may add all mathematicalness, depends essentially on getting each operation performed by *two methods*; the first, a roundabout one, which represents and registers, the conscious action of the mind during the process of discovery; the second, a short method which condenses the roundabout one, assists in stowing its results away in the memory, and facilitates the using of them subconsciously. (918)

The ordinary good citizen, whether of the unlearned or of the learned world, is fond of saying that you must not push any logical argument to extremes, or it lands you in some absurdity. Wherever mathematics touches, it instructs us to push everything to the extreme (in thought, in words), in order to be landed in absurdities; because only so can our premises be thoroughly tested. It is only by pushing in thought our tangent to infinity and our radius to the focus that we can properly understand our curve. The object of so pushing to the ultimatum is that we may understand the curve. When we understand it, we will discard altogether those fictitious entities, tangent and radius; *that* is for the mathematician, a matter of course. But the world imagines that the tangent or the radius—or both—are meant for lines of direction as to conduct; and is shocked, or argues to prove that such conduct would be unwise. (710)

The nature of mathematics

All the modern higher Mathematics is based on a Calculus of Operations, on Laws of Thought. All Mathematics, from the first, was so in reality; but the evolvers of the modern higher Calculus have known that it is so. Therefore elementary teachers who, at the present day, persist in thinking about Algebra and Arithmetic as dealing with Laws of Number, and about Geometry as dealing with Laws of Surface and Solid Content, are doing the best that in them lies to put their pupils on the wrong track for reaching in future any true understanding of the higher Algebras. Algebra deals not with Laws of Number, but with such laws of the human thinking machinery as have been discovered in the course of investigations on number. Plane Geometry deals with such Laws of Thought as have been discovered by men intent on finding out how to measure surface; and solid Geometry with such additional Laws of Thought as were discovered when men began to extend Geometry into three dimensions. The branch of Mathematics called Quaternions deals with such Laws of Thought as reveal themselves during the process of investigating the relations between n and $n + 1$ dimensions. The sooner pupils are made to see all Laws of Arithmetic as Laws of Thought, not of things, the simpler and more satisfactory will their future course be. (809)

The whole process of numeration depends on and expresses the power and the need of the human mind to merge a group of its ideal units of one stage into a single higher unit. It represents therefore the Law of Sacrifice. It embodies indeed both forms of that Law; the merging of lower units into higher; and the decomposition or breaking down of a higher unit into lower ones.

When we speak of sacrifice in this connection, we attach to it, of course, no idea of emotion or of tragedy; we refer only to the metaphysical conception of a law which underlies all our emotions and tragedies. The farmer goes forth casting his "rice seed on the waters," and his corn down to rot in the ground, "sowing in tears," but omitting the tears because he understands what he is doing, and knows he is not wasting the good seeds but preparing the crop of the future. We also in our sums, go through all the tragic drama of life, with the tragic element suspended by the presence of pure abstract comprehension.

All these, and several other conceptions which are essentially metaphysical, or spiritual, or psychological, are dealt with in Arithmetic and Algebra. . . . (793)

On multiplication

All mathematics would be simpler if we used a non-numerical phraseology, as we surely should do if we were less enslaved than we are by the convention of asserting that Arithmetic deals only with questions of number. Much confusion in the teaching of fractions is caused by popular mis-use of the words "to multiply" and "multiplication," which essentially belong to the domain of number, and have no meaning where the symbols involved represent unity, negation, or fraction. Some word should be chosen instead of "multiply" which is not associated with the idea of *number* or of *increase*; this word should be defined in the teacher's mind as "doing to the operand what, when done to Unity, produced the operator" and should be explained by him to the children in a sense consonant with that definition. What but confusion can come from telling children to multiply half a shilling by one-half, when the thing you mean them to do is to diminish the equivalent of sixpence down to threepence? What you want him to do to the half-shilling is something which, if done to the unit of thought, produces the conception one-half. The definition here given is valid for all values numerical, fractional, and logical. Whatever be the terminology chosen, it should lay the foundation for a habit of thinking of the symbol $\times$ as suggesting not the increase of things, but a mental action. This facilitates the future conception of reasoning by analogic process. (792)

The teacher should, in dealing with any branch of the operation called "multiplication", have in his mind this idea:— "Multiplication" means doing to the operand (multiplicand) what, if done to unity (1), would produce the operator (multiplier).

In dealing with "division" he should keep in his mind the idea:— "Division" means doing to the operand (dividend) what, if done to the operator (divisor), would produce unity (1). (811)

On algebra

I have pointed out how a question which seemed, on its own level, hopelessly unsolvable, is often solved at once by reference to an order of considerations higher than that involved in the question itself. This is especially the case when a *Law of Thought* is appealed to, to settle a discussion about *things*. Laws of things we cannot be said really to know, except in a fumbling and empirical manner; when we have true knowledge, it is because we have discovered the Law of Thought which presided at the Genesis of the Things. I wish to speak in all humility and reverence; but I cannot say less than I mean. We get clues, suggestions of Laws of Thought, by studying things; but whenever we truly *know*, what we know is a Law of Thought, which we have arrived at by discharging from our observation of particular finite things all that made them finite and particular. The elementary geometrician who first conceived the idea of the circle caught his suggestion from looking at things whose forms were approximately round; but, as soon as he had discovered the law of roundness within his own mind, he was able to express roundness in a new material, to state it generally (by scratching it on the sand) in a manner which afforded no clue to the objects from which the suggestion had come to him. And the law of circularity, thus formulated, was henceforth master within him, and governed his appreciation of things. He did not test his Ideal circle by comparing it with the sun or with an apple; on the contrary, he tested the circularity of a fruit by comparing it with his abstract or Ideal circle. His circle then was an *algebraization* of the round outline of the sun or moon, or of a fruit.

In the same way we discover a law of number, first, by thinking of some particular numbers; but as soon as we know the Law, we can state it *Algebraically*, *i.e.*, in a manner which conveys no information as to what were the particular numbers of which we happened to be thinking when we discovered it. The particular numbers suggested the law to our consciousness; they do not prove it to our reason. When once it has been suggested, it carries its own evidence, independently of particular numbers. And as soon as we have formulated a Law thus algebraically, it is henceforth master within us. Particular statements about number are referred to it; and our opinion as to the truth of those statements is controlled by it. For instance, the law that one number multiplied by a second always comes to the same result as is obtained by multiplying the second by the first was of course suggested to the consciousness first by the observation of some particular pairs of numbers; but it is not proved by reference to any special numbers; it is general and algebraic. And no student thoroughly understands it as a law of number until he understands it in its algebraic statement: $ab = ba$. As soon as he understands the algebraic statement, it becomes master of his thought. (444)

It is instructive to reflect that up to a few hundred years ago it was supposed that mental labour cannot be lightened by Mechanical Notations unless its subject is actually number. Questions of form and motion were still treated in the laborious round-about manner of which Euclid's Geometry is the type. Descartes, Newton, Leibnitz and others perceived that the principle by which our arabic notation diminishes labour is a principle not of number but of human thought; that economy of thought-force by diminishing strain on attention could be applied in dealing with form and motion. Babbage mentions a problem which the best older Geometers left obscure after four columns folio of verbal explanation; but which is made perfectly clear in four lines by Descartes' notation . . .

Algebraic notation had been supposed to be only applicable to number; to see it applied to form, puzzled people, and no doubt seemed unnatural and uncanny. But it is all settled now; everyone has accepted the idea that Algebra can be applied to form as well as number.

In this century, notation-thinking has invaded Chemistry, which before had been investigated only by empiric experiments and roundabout speculation, and taught by verbal explanations. And now people imagine that Algebra has something specially to do with number and form and chemistry, and that it can have nothing to do with other subjects. It would be just as true to suppose that a steam-engine can do no sort of work except the kinds it was first applied to. The steam engine is not an appendage of this or that kind of work, but an invention to supplement the weakness of human muscle. Algebraic notation is not an appendage of number or form, but an appliance intended to supplement man's limited power of attention. The original steam-engine can only do its own work; but once the principle was understood, it could be adapted to any kind of physical labour, by suitable adjustments; so with thought-instruments; the original Algebraic notation was, so to speak, entangled with the special Laws of Numbers; but when we understand its principle, we can omit from it the few actual numerical laws and apply it to any kind of investigation. Newton, Leibnitz, Descartes, were all investigating Laws of Thought: Degerando calls his book on Notations: "L'Art de Penser"; the newest chemical notation has been founded on that used in a book written by a non-chemist and entitled "Laws of Thought"; General Babbage has kindly allowed me to copy an unpublished paper in which his father describes the course of his investigations; one sentence in it is this: "I pursued with ardour the attempt to discover some of those laws of *human thought* which have led to our knowledge of the physical Sciences." (604)

The difficulty of mathematics

The manner in which that most logical of all text-books, Euclid, has been misused to induce illogical habits of mind, repeats itself in most departments of elementary Mathematics. A good instance of such misuse can be shown in connexion with the subject of equivalent fractions. Let us suppose that the teacher wishes to prove to the class that three-quarters of one is equal to one-quarter of three. He takes as his unit of thought the concept apple. He too often begins by stating a thing which is not true in itself: viz. that three-quarters of an apple is the same thing as a quarter of three apples. For the purpose with which the teacher's mind is occupied at the moment (the equivalence of fractions), the statement is true; it is as true as that, from the shop-keeper's point of view, twelve pence are a shilling. But the child, who has not yet been introduced to the conception of equivalence of fractions, is at a point of view quite different. To the child's imagination "great lots of grub" have a fascination quite independent of the size of his own share. To suggest (for instance) that a bun for one child is the same as a hundred buns for a hundred children would be to insult the pupil's understanding and his feelings—feelings which affect him all the more keenly that as yet he does not know how to express them. If he *could* believe you, what would be the meaning of school treats and social picnics? In proportion as you ultimately induce him to feel that one bun for a child is the same as a hundred buns for a hundred children, he will be, in future, the worse man, citizen, political economist; and the less fit to apply mathematics to problems in real forces. But we are speaking now of the effect of such statements on his intellectual processes, at the age when his sensations are as yet unwarped. The statement that a quarter of three apples is "the same as" or "equal to" three-quarters of an apple has brought prominently into the field of his imagination two pictures—an apple, and a group of three apples. The natural way of making an apple equal to three apples is to bring forward two more apples. It is the way things used to be made equal when he was learning addition; why not now? The teacher has therefore started him looking out for the two more apples; and as what is said next does not seem to be tending in that direction, his attention is distracted; he gives only half his mind to what is said. Thus he fails to get hold of one or more links in the chain of argument. Some children soon recover themselves, and attend to the rest of what is said. And as the teacher seems to be satisfied with the reasoning and to expect them to be so, they imagine they are so. This helps to form a habit of intellectual dishonesty, of confusing sham proof with real proof. The children who are either less sharp or more logical and thorough look longer for the non-appearing two apples, and miss more links of the chain of proof. These children feel that the whole demonstration has passed in that region where grown-ups conduct a self-satisfied mental life in which children cannot share. Thus are formed habits of "hopeless non-comprehension", perhaps even of "self-protecting and contemptuous non-attention". (812)

Another fruitful source of insanity is the irreverence of teachers for the process of Algebraization. Most children have more or less of that luminous transparency which enables the individual to perceive Laws of Thought reflected in the workings of his own soul. As was pointed out in a former Chapter, he who has thus perceived a Law of Thought may be (and often is) absolutely sure that he has seen a general Truth. Teachers, forgetting to discriminate between Laws of Thought and Laws of Things, assert that the individual is no standard for the community; and that *no* Law of Nature can be generalized from a single instance. Many intelligent young people are thus made ashamed of the divinest gift within them, the power of spontaneous perception of thought-laws; and destroy the faculty by not using it. In some, however, the tendency to Algebraization is too strong to be repressed. The lad or girl, being told that "common sense and logic oppose themselves to the practice of generalizing from a single instance", draws the conclusion that his teachers are the opponents of what he feels to be the most sacred portion of his mental life; therefore he sets up a course of life openly defiant of "common sense" as embodied in the advice of those around him. For some years he is only "eccentric" and perhaps "unmanageable"; but the strain is often too great for the brain to bear; he has never been taught to distinguish Laws of Thought from Laws of Things; he carries the practice of *generalizing from one instance* over into the domain of things; he reasons from analogy or jumps to conclusions from mere association of ideas; and falls into a hopeless tangle of delusions—victims of the wilful irreverence of his teachers, and his own fidelity to the great truths which they are trampling underfoot. (450)

A science of inspiration

The great beauty and clearness of French mathematics, which set in about the time of the Encyclopedie, culminated in 1855, in a work on Logic, and on the inspired Intellectual Faculties, by Gratry, in which he proved that the calculus of Newton and Leibnitz was a supra-logical procedure, and that geometric induction is essentially a *process of prayer*; by which he evidently means an appeal from the finite mind to the Infinite for light on finite concerns. He said that Logic (as ordinarily conceived in his day) had only feet, whereas, treated as he suggested, it would acquire wings. My husband in the previous year called attention (*Laws of Thought*, p. 4) to the distinction between mathematical induction and the kind of induction known to observers of physical fact. My husband did not use the words "inspiration" and "prayer" about the former, as Gratry did; as I said, he avoided calling the attention of the unlearned to his work by words familiar to them in connection with "religion"; but he described the process of mathematical induction in terms which should have shown to any educated person what he meant. He revelled in Gratry's book. I have never met with any other Englishman who had seen the mathematical portion till it was pointed out to him by me; though as a religious writer Gratry is well known. What has become of Gratry's influence on science? I have heard that he was threatened with excommunication. The sentence was not carried out: it would have called attention to his work.....

At this point our reasoning must leave the earth and rise for a moment on its wings. Those who can understand nothing which does not refer either to the hoarding of minerals or the cracking of nuts will not be able to follow us. Shut your eyes for a moment and turn your gaze within. Think what must have been the effect of the intense Hinduizing of three such men as Babbage, De Morgan and George Boole on the mathematical atmosphere of 1830–1865. What share had it in generating the Vector Analysis and the mathematics by which investigations in physical science are now conducted? (957)

In mathematics, however, much of the painful sense of futility is avoided by our knowing exactly what sacrifice is necessary for obtaining the special knowledge sought. Nearly nineteen centuries ago, a suggestion was made, which, when followed up, imports into the subject of sacrifice for Inspiration much of the exactness which robs mathematical sacrifice of its sting; and changes the sowing in tears recommended by piety from the condition of vague scattering of treasures at random, to that of well-calculated planting of good seed in appropriate soil. The suggestion occurs in a passage in the New Testament, on which much mystical comment has been written by theologians. It was pointed out to me long ago, by Professor Boole, as containing a psychological clue specially well worth following up. It seems to me to sum up and make available for purposes of practical guidance that theoretical knowledge about the nature of our own inspiration-receiving powers which we gain from mathematics generally.

"The wind bloweth where it listeth, and thou hearest the sound thereof, but canst not tell whence it cometh or whither it goeth: so is every one that is born of the Spirit." This evidently cannot refer to wind regarded as travelling in a straight line; we do know of a trade wind whence it comes and whither it goes. The path of the whirlwind was a matter of observation long before the Christian Era, and might have been suggested at any time to an intelligent observer by the motion of particles of dust. If the reader will look at the diagram on a dust-whirl, and suppose himself to be one of a cloud of gnats caught in the gust, or to be steering one of a flotilla of ships caught in a hurricane, he will see that the sentence contains both an admirable picture of a society on which has descended an inspiration to think on some new topic, and a canon of guidance for an individual who desires to find the main line of progress and to preserve his own sanity and safety. (727)

The only great hiatus, so far as I know, which exists in mathematical teaching between sound theory and the best actual practice, is of this kind: Mathematical induction is essentially inspirational work and should be done during periods when discipline is relaxed; periods of leisure, of re-creation from labour; during the Sabbath of the Inspirer, the Unity; ultimately even during sleep. Children are expected to do it to order, under the same conditions as those in which they do ordinary study; *i.e.*, in school, where discipline must not be relaxed; where time must be kept; where examinations are looming ahead. Thus bad habits are formed, which weaken that synthetic faculty on which true inspiration depends, and which cause the pupil to become a victim either to routine or to pseudo-inspiration. (1001)

"Necessity is the mother of invention." Is it not possible that the very limitation of our direct numerical faculties has forced mathematicians to create, unconsciously, the ladder on which man can safely climb to The Infinite? Is it not possible that mathematics is, after all, a science, not of number and quantity, but of the conditions under which man can make his progress towards unknown Truth uniform and safe, and can preserve himself from being seriously misled by the mistakes which he is sure to make on his way?

This question unavoidably presents itself, sooner or later, not, of course, to all mathematicians—that is to say, all users of mathematical formulae—but to all serious students of mathematical philosophy. It is in this century that it has taken concrete form and come into open expression in print; but it seems to me that it must always have been haunting the imaginations of the more subtle thinkers, ever since the first Abacus was invented as an aid in reckoning; ever since circles were first drawn by some inspired savage, by means, perhaps, of two thorns linked together by a strip of bark.

But more than I have yet indicated has dawned on the imagination of a few mathematicians (in this century certainly, and I have reason to think earlier).

Is not mathematics emphatically a miniature edition, so to speak, of the science of sane inspiration? Does it not contain the clue to an accurate distinction between the sane inspirations of genius and its aberrations? (703)

A mathematician once said, in allusion to some of his own discoveries, "There are some things that could never be done unless some men would consent to be ill in order to do them." He meant to designate by the word "ill" a condition not morbid in itself, a condition strictly physiological not pathological, a condition which resembles parturition rather than disease. It is normally attended by some symptom often associated with illness; and, without being disease, it is a condition favourable to the setting up of certain forms of disease unless special precautions are taken to prevent it. No great original work in abstract science, no great generalisation in any natural or physical science, is made, nor is any great work of art generated, without the author having gone through one of these singular phases. (1003)

Embracing ignorance

Arithmetic means dealing logically with facts which we know (about questions of number).

"Logically"; that is to say, in accordance with the "Logos" or hidden wisdom, *i.e.*, the laws of normal action of the human mind.

For instance, you are asked what will have to be paid for six pounds of sugar at 3d. a pound. You multiply the six by the three. That is not because of any property of sugar, or of the copper of which the pennies are made. You would have done the same if the thing bought had been starch or apples. You would have done just the same if the material had been tea at 3s. a pound. Moreover, you would have done just the same *kind* of action if you had been asked the price of seven pounds of tea at 2s. a pound. You do what you do under direction of the Logos or hidden wisdom. And this law of the Logos is made not by any King or Parliament, but by whoever or whatever created the human mind. . . . .

When people had only arithmetic and not algebra, they found out a surprising amount of things about numbers and quantities. But there remained problems which they very much needed to solve and could not. They had to guess the answer; and, of course, they usually guessed wrong. And I am inclined to think they disagreed. Each person, of course, thought his own guess was nearest to the truth. Probably they quarrelled, and got nervous and overstrained and miserable, and said things which hurt the feelings of their friends, and which they saw afterwards they had better not have said—things which threw no light on the problem, and only upset everybody's mind more than ever. I was not there, so I cannot tell you exactly what happened; but quarrelling and disagreeing and nerve-strain always do go on in such cases.

At last (at least I should suppose this is what happened) some man, or perhaps some woman, suddenly said: "How stupid we've all been! We have been dealing logically with all the facts we knew about this problem, except the most important fact of all, the fact of our own ignorance. Let us include that among the facts we have to be logical about, and see where we get to then. In this problem, besides the numbers which we do know, there is one which we do not know, and which we want to know. Instead of guessing whether we are to call it nine, or seven, or a hundred and twenty, or a thousand and fifty, let us agree to call it x, and let us always remember that x stands for the Unknown. Let us write x in among all our other numbers, and deal logically with it according to exactly the same laws as we deal with six, or nine, or a hundred, or a thousand."

As soon as this method was adopted, many difficulties which had been puzzling everybody fell to pieces like a Rupert's drop when you nip its tail, or disappeared like bats when the sun rises. Nobody knew where they had gone to, and I should think that nobody cared. The main fact was that they were no longer there to puzzle people. (1231)

This is only part of the essence of Algebra, which, as I told you, consists in preserving a constant, reverent, and conscientious awareness of our own ignorance. (1235)

Always remember that the use of algebra is to *free people from bondage*. For instance, in the case of number: Children do their numeration, their "carrying", in tens, because primitive man had nothing to do sums with but his ten fingers.

Many children grow superstitious, and think that you cannot carry except in tens; or that it is wrong to carry in anything but tens. The use of algebra is to free them from bondage to all this superstitious nonsense, and help them to see that the numbers would come just as right if we carried in eights or twelves or twenties. It is a little difficult to do this at first, because we are not accustomed to it; but algebra helps to get over our stiffness and set habits and to do numeration on any basis that suits the matter we are dealing with. (1239)

The essential element of Algebra: the habitual registration of the exact limits of one's knowledge, the incessant calling into consciousness of the fact of one's own ignorance, is the element which Boole's would-be interpreters have left out of his method. It is also the element which modern Theosophy omits in its interpretation of ancient Oriental Mind Science.

Men who wish to exploit other men fear nothing in logic or science except this element. They fear nothing in earth, heaven, or hell, so much as a public accustomed to realise exactly *how much has been proved, and where its own ignorance begins*. Exploiters fear this about equally, whether they call themselves priests, schoolmasters, college dons, political leaders, or organisers of syndicates and trusts. (1261)

Teapot algebra

Many people think that it is impossible to make Algebra about anything except number. This is a complete mistake. We make an Algebra whenever we arrange facts that we know round a centre which is a statement of what it is that we want to know and do not know; and then proceed to deal logically with all the statements, including the statement of our own ignorance.

Algebra can be made about anything which any human being wants to know about. Everybody ought to be able to make Algebras; and the sooner we begin the better. It is best to begin before we can talk; because, until we can talk, no one can get us into illogical habits; and it is advisable that good logic should get the start of bad.

If you have a baby brother, it would be a nice amusement for you to teach him to make Algebra when he is about ten months or a year old. And now I will tell you how to do it.

Sometimes a baby, when it sees a bright metal teapot, laughs and crows and wants to play with the baby reflected in the metal. It has learned, by what is called "empirical experience", that teapots are nice cool things to handle. Another baby, when it sees a bright teapot, turns its head away and screams, and will not be pacified while the teapot is near. It has learned by empirical experience, that teapots are nasty boiling hot things which burn one's fingers.

Now you will observe that both these babies have learnt by experience. Some people say that experience is the mother of Wisdom; but you see that both babies cannot be right; and, as a matter of fact, both are wrong. If they could talk, they might argue and quarrel for years; and vote; and write in the newspapers; and waste their own time and other people's money; each trying to prove he was right. But there is no wisdom to be got in that way. What a wise baby knows is that he *cannot tell* by the mere look of a teapot, whether it is hot or cold. The fact that is most prominent in his mind when he sees a teapot is the fact that *he does not know* whether it is hot or cold. He puts that fact along with the other fact: that he would very much like to play with the picture in the teapot supposing it would not burn his fingers; and he deals logically with both these facts; and comes to the wise conclusion that it would be best to go very cautiously and find out whether the teapot is hot, by putting his fingers near but not too near. That baby has begun his mathematical studies; and begun them at the right end. He has made an Algebra for himself. And the best wish one can make for his future is that he will go on doing the same for the rest of his life.

Perhaps the best way of teaching a baby Algebra would be to get him thoroughly accustomed to playing with a bright vessel of some kind when cold; then put another just like it on the table in front of him, one being filled with hot water. Let him play with the cold one; and show him that you do not wish him to play with the other. When he persists, as he probably will, let him find out for himself that the two things which look so alike have not exactly the same properties. Of course, you must take care that he does not hurt himself seriously. (1233)

SELECTED WRITINGS ON THE WORK OF GEORGE BOOLE

The laws of thought

My husband told me that when he was a lad of seventeen a thought struck him suddenly, which became the foundation of all his future discoveries. It was a flash of psychological insight into the conditions under which a mind most readily accumulates knowledge. Many young people have similar flashes of revelation as to the nature of their own mental powers; those to whom they occur often become distinguished in some branch of learning; but to no one individual does the revelation come with sufficient clearness to enable him to explain to others the true secret of his success. George Boole, poor and with little leisure for study, became known as a learned and original mathematician at an early age. From the first he connected his scrap of psychologic knowledge with sacred literature. For a few years he supposed himself to be convinced of the truth of "the Bible" as a whole, and even intended to take orders as a clergyman of the English Church. But by the help of a learned Jew in Lincoln he found out the true nature of the discovery which had dawned on him. This was that man's mind works by means of some mechanism which "functions normally towards Monism". Besides the information which reaches it from the external world, it receives knowledge direct from The Unseen every time it returns to the thought of Unity between any given elements (of fact or thought), after a period of tension on the contrast or antagonism between those same elements.

At this point all possibility of becoming a priest came to an end. George set to work to write a book (*The Laws of Thought*), in order to give to the world his great discovery. If he had stated it in words, he would have been entangled in an unseemly theological skirmish. He presented the truth to the learned, clothed in a veil so transparent that it is difficult to conceive how any human being could have been blinded by it; he proved that by the mere device of always writing the symbol 1 for whatever is the "Universe of Thought" for the time being, the whole cumbersome mechanism then known as "Logic" could be dispensed with. If you are thinking of sheep as divided into white and not-white, but x for 'white' and 1 for 'sheep'; if you are thinking of sheep as a portion of the animal kingdom, write z for 'sheep' and 1 for 'animals', and so on. Using this simple device, he proved that the most complicated examples given in any treatise on logic could be solved easily and mechanically by the ordinary processes of elementary algebra.

The academic world was enchanted. George visited Cambridge in 1855, a year after the publication, and was astonished, and at first gratified, at the cordiality of his reception. Herbert Spencer said that the book was "the greatest advance in Logic since Aristotle". George Boole said to me that neither Aristotle's Logic nor the Creed of Moses could have been enunciated unless the formula to which the Universities had now given the name of "Boole's Equation" had been, in some form or other, perfectly well known. George afterwards learned, to his great joy, that the same conception of the basis of Logic was held by Leibnitz, the contemporary of Newton. De Morgan, of course, understood the formula in its true sense; he was Boole's collaborator all along. Herbert Spencer, Jowett, and Leslie Ellis understood, I feel sure; and a few others. But nearly all the logicians and mathematicians ignored the statement that the book was meant to throw light *on the nature of the human mind*; and treated the formula entirely as a wonderful new method of reducing to logical order masses of evidence about external fact. Only think of it! The great English religious mind, which considers itself competent to preach *the Truth*, the only saving Truth, to all mankind; the great academic educational mind which is to improve Hindu culture off the face of the earth, fell into a trap which I believe would hardly have deceived a savage. My husband said to me that he believed he could never have made his discoveries if he had received a university education (as he at one time much wished to do, but was, fortunately, prevented by poverty). My after-experiences, among men who had been subjected to that process, incline me to think he was quite right in so believing. He was, as I said, a quiet student, gentle, timid, very conscientious, and averse to controversy; he could not face the theological animus which would be aroused by any attempt to explain himself in open words, nor did he feel it right to unsettle the superstitions of people evidently too stupid to take in reasonable truth; he went on to further researches. (951)

Boole's discoveries

The first of these discoveries he made in the year 1832 or 1833, at the age of seventeen. As he long afterwards told me, it flashed upon him suddenly one afternoon as he was crossing a field. It was the mystic secret, the open secret of all the ages; the same which has been known to all great Seers, Mystics, Religion-Founders from the dawn of History, viz:— that the mind of man is encased in a mechanism which, besides receiving impressions through what we call the senses, receives information also from some source, invisible and undefinable, the access to which opens whenever the mind, after a period of tension on the difference, contrast, or conflict between any elements of thought, turns to contemplate the *same* elements as united, or as forming parts of a unity. The highest form of this mental exercise is, of course, that recommended by certain religious moralists, i.e., when we wish for religious inspiration, to prepare for it by seeking the bond of brotherhood and friendship with the person with whom we have been at enmity. This form of reunion after separation is the highest in illuminating power; the preliminary tension, or sense of separateness, being more acute, and usually more prolonged, than where the sense of contrast is merely intellectual; and more of the nervous system is involved in personal hatred or anger than in a mere sense-perception, such as that the harebell is blue and the ragweed yellow. But a small and non-emotional effect of the kind indicated is produced when we turn to consider the qualities common to two plants or other objects, after we have been studying the respects in which they differ.....

This secret for inducing inspirations of new knowledge is embodied in Boole's Thought-Equation,

$$x + (\text{not-}x) = 1,$$

where 1 stands for the Unit of Thought for the time being. George Boole often tried to translate this Equation into words. But many circumstances combined to hinder his publishing any word-versions of it. In 1855, to his great joy, Père Gratry the Oratorian, formulated the truth which it expresses, in language on which George Boole felt he could not improve. Gratry made the Law of re-union after tension on contrast the basis of a method of procuring suggestions and revelations; a method as systematic and regular, and as scientific as any invention used by electricians for inducing currents. Gratry explains how he also used his method for the purpose of procuring sound and dreamless sleep and for making such sleep fertile of new thoughts. He prepared both for repose and for inspiration by fixing his mind on the properties common to things seemingly different, and about which he had been acquiring concrete knowledge by methods totally diverse. (789)

Now George Boole's second discovery was that of a *solvent* which, so to speak, soaks away and washes out from Arithmetic all questions of Number, and leaves the rest of it standing, complete, a perfect skeleton-picture of the mind's normal action when engaged in evolving the Science of which Number is the material. It thus stands clear as a contribution to the Science of human

Psychology, showing how minds of old worked when engaged in mastering difficulties.

Having thus reduced Algebra to its fibrous skeleton, he next compared it with that of Logic, and proved their complete homomorphism.

What I have called the solvent consists in an Equation the roots of which (or, in school-boy language, the answers to which) are 0 and 1. That Equation is

$$x^2 = x$$

or, otherwise

$$x \times (1 - x) = 0.$$

This equation is "satisfied" (*i.e.* turns out true), if either of the values 1 or 0 is substituted for x, but not if any *number* is substituted for x. An instance given in the Laws of Thought to illustrate the logical bearing of this equation is the following: From all the creatures in the world, select in your mind, sheep. Let the class-concept "sheep" be, for the moment, your unit of thought. From that unit, select the class of "horned things"; you thus get the concept "horned sheep", in contrast with a rejected background "sheep that are not horned". Your selection of "horned" from among sheep has really altered your concept "sheep", given to it clearer definition and a narrower scope. Now repeat on the class-concept, "horned sheep", the operation of selecting "horned things", this does not have any further effect; "horned sheep which are horned" are identically the same class as "horned sheep". On the other hand, if you apply the operation of selecting the horned individuals to the other class, "not horned sheep", the result is 0; for there are no horned things in that class.

Boole proved (to the entire satisfaction of the learned) that the so-called "dictum of Aristotle", (on which all Logic taught in European Universities up to his time had been founded) was a particular case of the equation,

$$x \times x = x.$$

The learned failed to seize the gently suggested moral; I will give it here in plain words. This is it: "Aristotle, by treating mind as a mechanism whose normal action is to swing between 1 and 0, created a system of Logic which has stood the test of two thousand years. Let us now discard that system, and learn, instead, how to use our minds as Aristotle used his, viz., in accordance with its Natural Laws. Let us dare, henceforth, to be every man his own Aristotle. Let us make methods in Logic, each for himself, as we go along, by treating each element of thought alternately as 1 and 0.

This is the *real* "Boole's Method", as practised by himself. To prove to the world the feasibility of his plan, George Boole applied it to create a new Formal Logic, a "Method", related to that hitherto used in Logic-classes, as a quartz-crushing machine is to a quarry-man's hammer. This method (known as Boole's Calculus) depends on treating his two Equations—The Mystic Law, or psychological equation,

$$x + (\text{not-}x) = 1$$

and the logical equation

$$x \times x = x$$

as simultaneous. Boole's logical Calculus is worked out between them. It is needlessly clumsy; and has been modified and much improved by such men as Jevons and Venn. None of their systems, however, concerns the ordinary business of life. Boole's real method: *Let us be everyone his own Aristotle*, concerns everybody, especially teachers of children.

After the publication of *The Laws of Thought*, George Boole wrote two text-books, assisted in bringing his grand conception down to the level of ordinary mathematical notation, by his enthusiastic pupils. He often longed to write still further on Psychology, which he was all the time studying devotedly, with the help of all the medical knowledge he could get. He was hindered from publishing anything further on the subject by not being able to clear up certain points; till, shortly before his death, I suggested to him that a certain well-known passage in the New Testament about the Wind, might refer to the course of a Circular Storm or Dust-whirl. After thinking it over for some time George Boole said: "Stick to that; you have got the clue now." He added that it would be worth my while to lay aside all other lines of inquiry and follow up the clue which I had found. He died shortly after this. (794)

No amount of skill in using a mathematical notation throws any light on the conditions under which it was generated. A man may use a scientific notation with consummate skill and yet know no more of the mode in which it was generated than the boy who turns the handle of a machine need know of the nature of the investigations which presided over its construction. Many so-called mathematicians are so unawake to the true nature of the machine which they are manipulating that they are hardly aware that there is anything to learn. I, who lived nine years with George Boole while he was collaborating with De Morgan, know—that I do not know. I do as George Boole and De Morgan did: I bow my head in reverent thankfulness to that mysterious East, whence come to us wafts of some transcendent power the nature of which we ourselves can hardly state in words. When materialists on one side, and theosophists on the other, agree in assuring the public that the great structure of European science has been created *without reference to psychic lore*, I feel that I owe it to the cause of truth to say that I differ from these persons, not as to the truth or value or originality of this or that idea or statement, but as to the contents of books which they have apparently never read, about the genesis of notations which they can only use mechanically (if they can even do that much), whereas I have discussed the details of some of these with the originator before their form was finally fixed. (961)

I was asked to translate the Life-Laws, or Laws of Thought (of Gratry and Boole), out of the language of the Modern Calculus into that of the simple geometry used of old. To exhibit the most important law I had to use a diagram in shape like a capital V. I had to draw the thing open end upwards, and afterwards shut end upwards. I noticed that my stupider pupils were sometimes confused by the first V remaining on the paper; I wished I could pick it bodily up and reverse it, to show that the same thing was seen alternately in the two aspects. A pair of compasses, open at the Freemason's Angle, would have answered exactly, except that it did not show the *process of becoming*, as a drawing made before the pupil does. What was wanted was, a thing that would first suggest *growing from one into two*, and that I could then turn upside down. In fact, a natural forked stick. While thinking this over, I suddenly remembered seeing a miner doing something (in my childhood) with a stick the very shape I wanted. This excited my curiosity, and I went to a country parish where I had heard there was a lady dowser, and asked her to show me her magic.

The whole process is this. The dowser voluntarily goes through, in dumb show, half of a logic lesson on the first Law of Thought. While he is doing so, his arms begin to tingle; and, as he passes over ore, or running water, his hands automatically complete the lesson by reversing the rod. The dowser, who knows no logic, does (one half by tradition and the other half automatically) what I do on purpose to teach the first Law of Thought. (394)

On singular solutions

Draw a circle with a radius of an inch, and then ink it over so as to distinguish it from all the others which you are going to draw.

Then close your compasses to a quarter of an inch. Put the point anywhere on our ink line, and draw a circle in pencil. Put your point down anywhere else on the ink line, and draw another circle. Repeat again and again, till you have a dozen circles, or a hundred, or indeed any number that you like. Any point in the ink line is the centre of one such possible circle; and, as the number of points in the circumference is infinite, there is no limit to the number of such circles, except what may be imposed by the limitations of your patience or your eyesight.

Well, all these little pencil circles, actual or possible, belong to what is called a "family". Any "family" of curves has certain laws to which all the members are subject, and which are the laws of their being. What are the laws of this family's being? "Every member of it has its centre on the ink line, and is exactly half an inch across", you say. No you are wrong. All the members *that ever you saw yet* are indeed subject to those conditions which you mention; and if you thought of them as isolated individuals you might never find out that there was anything more to say about them. But when a mathematician expresses in strict mathematical language the law which binds those circles into a family, and states their connection with each other, he finds (to his great surprise, if it is his first experience of the kind) indications of the existence, somewhere or other in space, of two other curves belonging to the family, which have a larger radius and whose centres are not situated on the ink line. On inquiring further, he finds out whereabouts in space these larger circles are.

Draw two circles, whose centre coincides with that of the ink circle, and whose radius is in the one case three quarters in the other five quarters, of an inch; these two circles belong to the same family as the small ones. (The ink circle does not.) These larger circles are instances of what mathematicians call "Singular Solutions". Ordinary people call them "exceptions". Philosophers call them "typical cases". Impatient people call them "anomalies", and do their best to weed them out, so as to reduce everything to the domain of what they please to call "law"—by which they mean such law as they can understand.

A Singular Solution may be located anywhere in space; often it is a long way off from its family, and of very different shape. In this particular case, and many others, it is of the same shape as the more commonplace members, and in contact with every one of them. Sometimes the law of a family has one Singular Solution; sometimes more than one. (252)

The study of the phenomenon of the Messianic Seerhood became the absorbing interest of George Boole's later years. The chapter on Singular Solutions in his text-book of differential Equations contains much genuine metaphysical truth expressed in mathematical terminology. He made a special study of the sermons of the strange preacher, F. D. Maurice; and in the last year of his life wrote: "I have made out what puts the whole subject of Singular Solutions into a state of Unity." What strange pulsation of thought must have thrilled from the mind of the great Seer of Messianic Singular Solution, into that of the logician of the Unity, before those words were written, who can guess? The MS. on which George Boole was engaged at the time lies undeciphered in the archives of the Royal Society; no mere mathematician can understand it: and no theologian cares to try. Perhaps it may be deciphered yet, when further progress has been made in knowledge of the correspondence between the normal action of the human mind and its automatic expression by means of notations. (444)

The subject of Singular Solutions had a curious fascination for my husband, then and to the end of his life. The chapter of the text-book which refers to them was written in a sort of religious rapture . . . (29)

Charles Babbage is chiefly known to the world as the inventor of a machine intended to spare the labour of calculating numerical series. In the course of its construction, he had to make a thorough study of the Laws of natural sequence, so far as these are embodied in processes of successive additions. But the man who would think of undertaking such a task at all was sure to see the importance of saturating himself with a further knowledge of Nature's series. He investigated not only those mathematical series the equations of which are known, and which underlie such natural curves as planet-paths, lines of refraction, etc.: but also those forms of Natural sequence the mathematical expressions of which have not yet been ascertained, such as geologic changes, the development of plants and animals, etc. He had nothing to lose or gain by any conclusion to which he might be led; he had one all-absorbing end in view, the perfecting of his machine; and, for that object, it mattered nothing what the Laws of Nature should turn out to be; the one desideratum was that he, Babbage, should know what they were, and embody them truly in the construction of his cogs and wheels. One of the facts which he discovered was this: For one series (numerical or phenomenal) which goes on uniformly, there are an almost infinite number which either sooner or later have interruptions of Singular Terms. His arguments cannot be fully entered on here; but the main result is this: Whoever adduces the inflexibility of Law to prove the improbability of Miracle, only proves that he does not understand the connection between Law and Phenomena. No miracle, Mr. Babbage considers, could be *a priori*, so improbable as it is that man should learn the true law of any sequence by observing an *uninterrupted* series of phenomena. (424)

Various mathematical studies have thrown a vivid light on the phenomenon of inversion; among them is one by Babbage, published in 1837, under the title *Ninth Bridgewater Treatise*. Lines and shapes in three dimensions which, in themselves, are continuous and normal, produce, if their shadows are cast on flat paper, all manner of anomalies, inversions, and irregularities. It is inferred that normal facts in higher dimensions might, when reflected on the human three-dimensional consciousness, account for anomalies and apparently miraculous conditions. The experiment was therefore tried of inverting certain relations in the equations of regular curves and families of curves. A most interesting result followed. The inverted equation to a curve sometimes gives what is called a "singular point", related to the original curve almost as a "rogue elephant" is to his tribe; standing aloof, in no relation visible to the eye with any other point on the curve. On the other hand, the invert equation to a family of curves may produce what is called an "envelope", *i.e.* a curve in contact with every other member of the family. These anomalous points or curves, worked out in invert equations, are called in mathematical terminology, *singular solutions.* Numa Hartog, the young Jew whose brilliant achievement in mathematics was the occasion of opening university honours to his co-religionists, said to me, in a puzzled kind of way, that the chapter on singular solutions in my husband's *Treatise on Differential Equations*, "does not read like a chapter of an ordinary text-book". It would be strange if it did, seeing that it resulted from a devout study of the mystic literature of all ages, from Isaiah's prophecies to Tennyson's "St. Agnes' Eve". My husband, in the last years of his life, devoted much time to the study of Frederick

Denison Maurice, the theological reformer, and, not long before his death, wrote: "I have made out what puts the whole subject of Singular Solutions into a State of Unity." Following up clues given by him at that period some extra-academic mathematicians have worked on the problem so near his heart throughout his life, viz.: Given a child of any anomalous type, how should his education be directed towards forming in him personal habits tending towards constructive genius rather than perverse destructiveness; towards illuminating synthesis rather than brilliant paradox; towards useful originality rather than vicious curiosity; towards a reverent and sparing use of pleasures, towards renunciation of that for which other men seek, rather than towards audacious snatching at what they feel to be evil?

When I have tried to call the attention of heads of schools to the dangers to which children of exceptional type are exposed under existing schemes of moral and religious education, I have been met by replies to the effect that a school must be organised to suit the majority, and that exceptional children must take their chance in the general melee—a dangerous doctrine in itself, it seems to me, because genius has great influence, for good or ill, over the masses of average men and women. But besides this, it is thought by several experienced persons that the scheme of mental habit suggested for the protection of genius from its special mental and moral dangers would in reality be good, not bad, for the development and stamina of average children.

We are often told by publishers and editors that the public is not interested in speculation of this kind. But if England takes no interest in the question whether its young men of abnormal genius shall lay themselves on the altar of National Reform or rot away in mere phosphorescent decadence, why was Oscar Wilde condemned to prison, and why do we perform religious services in honour of Jesus of Nazareth? (1380)

The last few paragraphs of the Investigation into the Laws of Thought *by George Boole:*

Those who have maintained that the position of mathematics is . . . . a fundamental one, have drawn one of their strongest arguments from the actual constitution of things. The material frame is subject in all its parts to the relations of number. All dynamical, chemical, electrical, thermal, actions, seem not only to be measurable in themselves, but to be connected with each other, even to the extent of mutual convertability, by numerical relations of a perfectly definite kind. But the opinion in question seems to me to rest upon a deeper basis than this. The laws of thought, in all its processes of conception and of reasoning, in all those operations of which language is the expression or the instrument, are of the same kind as are the laws of the acknowledged processes of Mathematics. It is not contended that it is necessary for us to acquaint ourselves with those laws in order to think coherently, or, in the ordinary sense of the terms, to reason well. Men draw inferences without any consciousness of those elements upon which the entire procedure depends. Still less is it desired to exalt the reasoning faculty over the faculties of observation, of reflection and of judgement. But upon the very ground that human thought, traced to its ultimate elements, reveals itself in mathematical forms, we have a presumption that the mathematical sciences occupy, by the constitution of our nature, a fundamental place in human knowledge, and that no system of human culture can be complete or fundamental, which altogether neglects them. But the very same class of consideration shows with equal force the error of those who regard the study of Mathematics, and of their applications, as a sufficient basis either of knowledge or of discipline If the constitution of the material frame is mathematical, it is not merely so. If the mind, in its capacity of formal reasoning, obeys, whether consciously or unconsciously, mathematical laws, it claims through its other capacities of sentiment and action, through its perceptions of beauty and of more fitness, through its deep springs of emotion and affection, to hold relation to a different order of things. There is, moreover, a breadth of intellectual vision, a power of sympathy with truth in all its manifestations, which is not measured by the force and subtlety of the dialectic faculty. Even the revelation of the material universe in its boundless magnitude, and pervading order, and the constancy of law, is not necessarily the most fully comprehended by him who has traced with minutest accuracy the steps of the great demonstration. And if we embrace in our survey the interests and duties of life, how little do any processes of mere ratiocination enable us to comprehend the weightier questions that they present! As truly, therefore, as the cultivation of the mathematical or deductive faculty is a part of intellectual discipline, so truly is it only a part. The prejudice which would either banish or make supreme any one department of knowledge or faculty of mind, betrays not only error of judgement, but a defect of that intellectual modesty which is inseparable from a pure deviation of truth. It assumes the office of criticising a constitution of things which no human appointment has established, or can annul. It sets aside the ancient and just conception of truth as one though manifold. Much of this error, as actually existent among us, seems due to the special and isolated character of scientific teaching—which character, in its turn, tends to foster. The study of philosophy, notwithstanding a few marked instances of exception, has failed to keep pace with the advance of the several departments of knowledge, whose mutual relations it is its province to determine. It is impossible, however, not to contemplate the particular evil in question as part of a larger system, and connect it with the too prevalent view of knowledge as a merely secular thing, and with the undue predominance, already adverted to, of those motives, legitimate within their proper limits, which are founded upon a regard to its secular advantages. In the extreme case it is not difficult to see that the continued operations of such motives, uncontrolled by any higher principles of action, uncorrected by the personal influence of higher minds, must tend to lower the standard of thought in reference to the objects of knowledge, and to render void and ineffectual whatsoever elements of a noble faith may still survive. And ever in proportion as these conditions are realised must the same effects follow. Hence, perhaps, it is that we sometimes find juster conceptions of the unity, the vital connexion, and the subordination to a moral purpose, of the different parts of Truth, among those who acknowledge nothing higher than the changing aspect of collective humanity, than among those who profess an intellectual allegiance to the Father of Lights. But these are questions which cannot further be pursued here. To some they will appear foreign to the professed design of this work. But the consideration of them has arisen naturally, either out of the speculations which that design involved, or in the course of reading and reflection which seemed necessary to its accomplishment.

APPENDIX

Some people

ARCHER, MRS.

Friend of M.B., wife of the translator of Ibsen's plays. Developed a nerve-training method of alternate tension and relaxation—adapted by M.B. in her arithmetic lessons with children.

BABBAGE, CHARLES. (1792–1871)

Friend of Everests, brilliant polymath who devised mechanical computer and anticipated operational research. As a student he formed the Analytic Society with Herschel and Peacock to reform mathematics at Cambridge and "to do their best to leave the world wiser than they found it".

BOOLE, GEORGE. (1815–1864)

Husband of M.B., son of a shopkeeper, taught in his own school until a pamphlet on logic brought him encouragement from Morgan and a move to Cork. Brilliant and original mathematician—pure mathematics was invented by Boole according to Russell—essence of his ideas not understood by others according to M.B. He also made fundamental contributions to the theories of differential equations, finite differences and invariants.

BOOLE, LUCY.

Fourth daughter of M.B., became professor of chemistry in London, shared house with M.B. in Notting Hill.

BOSE, JAGADIO. (1858–1937)

Indian physicist and plant physiologist. He gave a lecture at the Royal Institution which impressed M.B. who corresponded with him and was stimulated to study Indian thought as both her uncle and De Morgan had done.

BOULANGER, NICHOLAS.

18th century writer whom M.B. claimed as precursor of Gratry and Boole. He had written of the misuse of special gifts and powers; M.B. was to write about the dangers of influencing young minds.

CLIFFORD, WILLIAM. (1845–1879)

Mathematician, made important and original contribution to geometry, author of *The common sense of the exact sciences*, professor at University College, London. A highly strung and athletic man, he was mentioned by M.B. as an example of a dominating teacher personality.

DE MERICOURT, VIRGINIA

Pseudonym used by M.B. for regular articles in various journals.

DE MORGAN, AUGUSTUS. (1806–1871)

Friend of the Everests, mathematician noted for his work on logic and for his interest in teaching, was influential in securing G.B. a post at Cork, fought for religious freedom at the University of London.

DEPLACE.

The village schoolmaster at Poissy who taught M.B. as a child.

EVEREST, SIR GEORGE. (1790–1866)

Uncle to M.B., surveyor and geographer, travelled and worked in India—Mt. Everest named after he made the first survey of it—interested in Indian mathematics and science.

EVEREST, REV. GEORGE

Brother of M.B., went to Canada, corresponded regularly with M.B.

EVEREST, REV. THOMAS

Father of M.B., family said to be of Hugenot stock, treated by Hahnemann and helped in his parish by M.B. on family's return from France.

GRATRY, FATHER.

French priest of the order of Oratorians. His book on logic published in 1855 was avidly read by G.B. and M.B. who found similar ideas to their own in his writings. M.B. developed his notion of suspending conscious thought to tap unknown powers—"making a silence in the soul".

HAHNEMANN, SAMUEL. (1755–1843)

Founder of homeopathic medicine, believed that diseases can be cured by drugs which produce similar symptoms, successfully treated father of M.B.

HERSCHEL, JOHN. (1792–1871)

Friend of Babbage and the Everests, astronomer, mathematician—he introduced the notation $\sin^{-1}x$, chemist—he developed photography and first referred to photographic images as positive and negative, translator of the Iliad and, like Newton, Master of the Mint in his later years.

HINTON, JAMES. (1822–1875)

Friend of father of M.B., an ear-surgeon and author of popularising works on philosophy and science. M.B. was his secretary for a few years.

JAMES, WILLIAM. (1842–1910)

Famous psychologist and philosopher, corresponded with M.B. with whom he shared an interest in progressive education and whose ideas he valued.

MARKS, DAVID.

Jewish preacher who had influenced G.B. and De Morgan. Through his work, M.B. became interested in Hebrew ritual and various current Jewish problems.

MAURICE, FREDERICK. (1805–1872)

Theologian, founded Christian Socialist movement, influenced many 19th century thinkers, helped to found Queen's College, London. His writings had interested G.B.

MORANT, ROBERT. (1863–1920)

Civil servant, architect of 1902 Education Act, knew and shared ideas with M.B.

PEACOCK, GEORGE. (1791–1886)

Fellow student of Babbage and Herschel, became Dean of Ely, wrote an influential textbook on algebra which aided the development of algebra as an abstract system.

PERRY, JOHN.

Mathematician, initiated reform in teaching by a famous address to the British Association in 1901 and a book on mathematics teaching, stimulating M.B. to write herself.

RYALL, JOHN.

Uncle of M.B., friend of G.B. to whom *Laws of Thought* was dedicated, professor of classics at Cork.

STOTT, ALICIA.

Third daughter of M.B., a mathematician who found elegant construction for the Archimedean solids and studied four-dimensional figures.

SOMERVELL, ARTHUR.

Friend of M.B., influential H.M. Inspector for music.

SOMERVELL, EDITH.

Friend of M.B., wife of Arthur S., author of *A rhythmic approach to mathematics*, a book about curve-stitching which had been introduced by M.B.

TAYLOR, MARGARET.

Second daughter of M.B., her son Sir Geoffrey Taylor is an eminent contemporary mathematical physicist.

VOYNICH, ETHEL.

Fifth daughter of M.B., became a novelist.

WEDGWOOD, THOMAS.

Son of the famous potter, friend of Coleridge, pupil of Erasmus Darwin. An invalid, he devoted his life to his brother's children and the study of education, in particular of the emotions; his thoughts strongly influenced M.B.

Some books

E. M. COBHAM, *Mary Everest Boole—a memoir*, C. W. Daniel, 1951.
A brief biography and account of her work, this is a useful companion volume by the editor of the collected works.

E. T. BELL, *Men of mathematics, Vol.* 2., Pelican, 1953.
Includes a short account of the life and work of George Boole. The author emphasises the early years of struggle, but then suggests—too facilely—that the marriage to the niece of the professor of classics represented a "subconscious striving for social respectability". There is a brief sketch of what has come to be known as Boolean algebra.

G. BOOLE, *An investigation of the laws of thought*, Dover reprint.
"The design of the following treatise is to investigate the fundamental laws of those operations of the mind by which reasoning is performed . . ." And he meant just that.

M. BOOLE, *Logic taught by love*, 1890.
Subtitled "rhythm in nature and in education" this book was based on a series of articles published in various Jewish journals in England and the U.S.A.

M. BOOLE, *The mathematical psychology of Gratry and Boole*, 1897.
An influential work which attempted a psychological interpretation of Boole's work and its relation to the thought of Gratry.

M. BOOLE, *Lectures on the logic of arithmetic*, 1903.
Addressed to young teachers, this is the book which contained specimen lessons and described the method of making mind-pictures.

M. BOOLE, *The preparation of the child for science*, 1904.
An attempt to explain to parents and teachers the foundations in preliminary experiences on which the learning of science and mathematics depends. The book anticipated many other thinkers and was particularly influential in the U.S.A.

M. BOOLE, *Philosophy and fun of algebra*, 1909.
An attempt to spell out more simply some of the themes explored in her previous books. In particular the process of algebraisation is described and discussed in some detail.

M. BOOLE, *The forging of passion into power*, 1910.
Those who may be attracted by the title will not be disappointed by this profound but neglected work. It explores the nature of unconscious processes and relates the insights acquired to problems of teaching. A powerful and prophetic book.

M. BOOLE, *Collected works, Vols.* 1–4, C. W. Daniel, 1931.
The publisher and his wife were close friends of Mary Boole. The collected works were distributed by the Open Court Company, U.S.A., at one time—they are now out-of-print.

C. GATTEGNO, *The psychology of the adolescent*, Educational Explorers, Reading, 1962.

C. GATTEGNO, *For the teaching of mathematics*, vols. 1–3, Educational Explorers, Reading, 1963,

C. GATTEGNO, *What we owe children*, Kegan Paul, 1970.
Perhaps the only contemporary writer on mathematical education who speaks the same language as Mary Boole—though very much in a voice of his own.

R. RHEES (ed.), *Studies in logic and probability*, 1952.
Various papers about George Boole with some of his own writings.

G. SPENCER BROWN, *Laws of form*, 1969.
"The theme of this book is that a universe comes into being when a space is severed or taken apart." An unusual contemporary account of a logical calculus and the process of mathematical creation.